SMP interact

T2

Teacher's guide to Book T2

CAMBRIDGE
UNIVERSITY PRESS

PUBLISHED BY THE PRESS SYNDICATE OF THE UNIVERSITY OF CAMBRIDGE
The Pitt Building, Trumpington Street, Cambridge, United Kingdom

CAMBRIDGE UNIVERSITY PRESS
The Edinburgh Building, Cambridge CB2 2RU, UK
40 West 20th Street, New York, NY 10011-4211, USA
477 Williamstown Road, Port Melbourne, VIC 3207, Australia
Ruiz de Alarcón 13, 28014 Madrid, Spain
Dock House, The Waterfront, Cape Town 8001, South Africa

http://www.cambridge.org

First published 2002

Printed in the United Kingdom at the University Press, Cambridge
Typeface Minion *System* QuarkXPress®

A catalogue record for this book is available from the British Library

ISBN 0 521 78541 3 paperback

Typesetting and technical illustrations by The School Mathematics Project
Cover image © Image Bank/Antonio Rosario
Cover design by Angela Ashton

Contents

The following people contributed to the writing of the SMP Interact key stage 3 materials.

Ben Alldred
Juliette Baldwin
Simon Baxter
Gill Beeney
Roger Beeney
Roger Bentote
Sue Briggs
David Cassell
Ian Edney
Steve Feller
Rose Flower
John Gardiner
Bob Hartman
Spencer Instone
Liz Jackson
Pamela Leon
John Ling
Carole Martin
Peter Moody
Lorna Mulhern
Mary Pardoe
Peter Ransom
Paul Scruton
Richard Sharpe
Susan Shilton
Caroline Starkey
Liz Stewart
Pam Turner
Biff Vernon
Jo Waddingham
Nigel Webb
Heather West

Others, too numerous to mention individually, gave valuable advice, particularly by commenting on and trialling draft materials.

Editorial team: David Cassell, Spencer Instone, John Ling, Mary Pardoe, Paul Scruton, Susan Shilton

Project Administrator: Ann White

Design: Pam Alford, Melanie Bull, Nicky Lake, Tiffany Passmore, Martin Smith

Project support: Carol Cole, Pam Keetch, Jane Seaton, Cathy Syred

Special thanks go to Colin Goldsmith.

Introduction

What is distinctive about *SMP Interact*?

SMP Interact sets out to help teachers use a variety of teaching approaches in order to stimulate pupils and foster their understanding and enjoyment of mathematics.

A central place is given to discussion and other interactive work. Through discussion with the whole class you can find out about pupils' prior understanding when beginning a topic, can check on their progress and can draw ideas together as work comes to an end. Working interactively on some topics in small groups gives pupils, including the less confident, a chance to clarify and justify their own ideas and to build on, or raise objections to, suggestions put forward by others.

Questions that promote effective discussion and activities well suited to group work occur throughout the material.

SMP Interact has benefited from extensive and successful trialling in a variety of schools. The practical suggestions contained in the teacher's guides are based on teachers' experiences, often expressed in their own words.

Who are Books T1 to 3 for?

Books T1 to 3 follow on from Books 1 and N and cover national curriculum levels up to 5.

How are the pupils' books intended to be used?

The pupils' books are a resource which can and should be used flexibly. They are not for pupils to work through individually at their own pace. Many of the activities are designed for class or group discussion.

Activities intended to be led by the teacher are shown by a solid strip at the edge of the pupil's page, and a corresponding strip in the margin of the teacher's guide, where they are fully described.

A broken strip at the edge of the page shows an activity or question in the pupil's book that is likely to need teacher intervention and support.

Where the writers have a particular way of working in mind, this is stated (for example, 'for two or more people').

Where there is no indication otherwise, the material is suitable for pupils working on their own.

Starred questions (for example, *C7) are more challenging.

What use is made of software?

Points at which software (on a computer or a graphic calculator) can be used to provide effective support for the work are indicated by these symbols, referring to a spreadsheet, graph plotter or dynamic geometry package respectively. Other suggestions for software support can be found on the SMP's website: www.smpmaths.org.uk

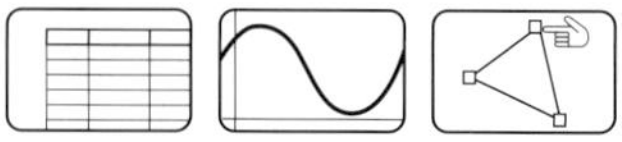

How is the attainment of pupils assessed?

The interactive class sessions provide much feedback to the teacher about pupils' levels of understanding.

Each unit of work begins with a statement of the key learning objectives and finishes with questions for self-assessment ('What progress have you made?'). The latter can be incorporated into a running record of progress.

Revision questions are included in periodic reviews in the pupil's book.

What will pupils do for homework?

The practice booklets may be used for homework.

Often a homework can consist of preparatory or follow-up work to an activity in the main pupil's book.

Answers to questions on resource sheets

For reasons of economy, where pupils have to write their responses on a resource sheet the answers are not always shown in this guide. For convenience in marking you could put the correct responses on a spare copy of each sheet and add it to a file for future use.

General guidance on teaching approaches

Getting everyone involved

When you are introducing a new idea or extending an already familiar topic, it is important to get as many pupils as possible actively engaged.

Posing key questions A powerful technique for achieving this is to pose one or two key questions, perhaps in the form of a novel problem to be solved. Ask pupils, working in pairs or small groups, to think about the question and try to come up with an answer.

When everyone has had time to work seriously at the problem (have a further question ready for the faster ones), you can then ask for answers, without at this stage revealing whether they are right or wrong (so that pupils have to keep thinking!). You could ask pupils to comment on each other's answers.

Open tasks Open tasks and questions are often good for getting pupils to think, or thinking more deeply. For example, 'Working in groups of three or four, make up three questions which can be solved using the technique we have just been learning. Try to make your questions as varied as possible.'

Questioning skills

Questioning with the whole class If your questions to the class are always closed, and you reward the first correct response you get, then you have no way of telling whether other pupils knew the correct answer or whether they had thought about the question at all. It is better to try to get as many pupils as possible to engage with the question, so do not at first say whether an answer is right or wrong. You could ask a pupil how they got their answer, or you could ask a second pupil how they think the first one got their answer.

Working in groups

Types of group work Group work may be small scale or large scale. In small scale group work, pupils are asked to work in pairs or small groups for a short while, perhaps to come up with a solution to a novel type of problem before their suggestions are compared. In large scale group work, pupils carry out in groups a substantial task such as planning a statistical inquiry or designing a poster to get over the essential idea of the topic they have just been studying.

Organising the groups Group size is important. Groups of more than four or five can lead to some pupils making little or no contribution.

For some activities, you may want pupils to work unassisted. But for many, your own contribution will be vital. Then it is generally more effective if, once you are sure that every group has got started, you work intensively with each group in turn.

After the group work One way to help pupils feel that they have worked effectively is to get them to report their findings to the whole class. This may be done in a number of different ways. One pupil from each group could report back. Or you could question each group in turn. Or each group could make a poster showing their results.

Managing discussion

Discussion, whether in a whole-class or group setting, has a vital role to play in developing pupils' understanding. It is most fruitful in an atmosphere where pupils know their contributions are valued and are not always judged in terms of immediate correctness. It needs careful management for it to be effective and teachers are often worried that it will get out of hand. Here are a few common worries, and ways of dealing with them.

What if ... '... the group is not used to discussion?'

- Allow time for pupils to work first on the problem individually or in small groups, then they will all have ideas to contribute.

'... everyone tries to talk at once?'

- Set clear rules. For example, pupils raise their hands and you write their name on the board before they can speak.

'... a few pupils dominate whole-class discussion?'

- Precede any class discussion with small-group discussion and nominate the pupils who will feed back to the class.

'... one pupil reaches the end point of a discussion immediately?'

- Tell them that the rest of the group need to be convinced and ask the pupil to convince the rest of the group.
- Accept the suggestion and ask the rest of the group to comment on it.

1 Number bites

These activities are intended as a resource to be dipped into throughout this book and are not meant to be worked through as a single unit.

These 27 activities practise various number skills covered in earlier SMP Interact materials. A wide range of activities is included: written questions, oral work, games, puzzles and investigations. Many of the activities can be used in a variety of ways, as class activities, group work or individual tasks in line with the range of teaching styles suggested at the start of this book. It is expected that the wide range of activities may suggest to teachers similar activities they could design themselves.

All the activities are expected to be done without a calculator.

The activities are grouped into whole numbers, fractions and decimals, and the order of the activities roughly represents an order of progression. Later units in this book rely on some of the skills practised here; these may therefore provide useful revision practice for weaker pupils immediately before starting these units.

Essential
Sheets 218 and 219

Whole numbers (p 4)

W3 Closest neighbour maze

Sheet 218

W6 Sweets

T

◊ This is intended for oral questions based on the prices of the sweets. Examples of possible questions asked are

- How much do 9 Sizzlers cost?
- How much do a Tiger and a box of Mint Chocs cost?
- How much more does a Tiger bar cost than a Nutty bar?

- How much change do I get from £1 if I buy a box of Mint Chocs?
- I have 50p to spend. How many liquorice laces can I buy?

W8 Menu

T

The menu is intended as a stimulus for oral questions, though some pupils may need to use pencil and paper. Examples of questions are:

- How much is a large hamburger with regular fries?
- How much dearer is a large fruit pie than a regular one?
- How much do you save by buying the big value hamburger meal rather than the separate items in it?

Fractions (p 8)

F3 Fraction maze

Sheet 219

Decimals (p 9)

D3 'One tenth more' bingo

T

Read out numbers which are 0.1 less than those in the pupil's book but include other numbers as well. The game is easily varied to 'one tenth less' or 'two tenths more', and so on.

D6 Rodents

T

Pupils can be asked questions orally or in written form. Some suggestions are:

- What is the total length of rodent 1?
- Which rodents have a longer tail than body?
- How much longer is rodent 6's tail than its body?
- How much longer in total is rodent 1 than rodent 2?

Whole numbers (p 4)

W1 Cheque it out

1 Six thousand and eighty pounds

2 Five thousand four hundred and sixty pounds

3 Ten thousand two hundred pounds

4 Two hundred thousand pounds

5 Fifty thousand and fifty pounds

W2 Hidden places

1 BERLIN 2 MADRID

3 NEW YORK 4 MOSCOW

W3 Closest neighbour maze

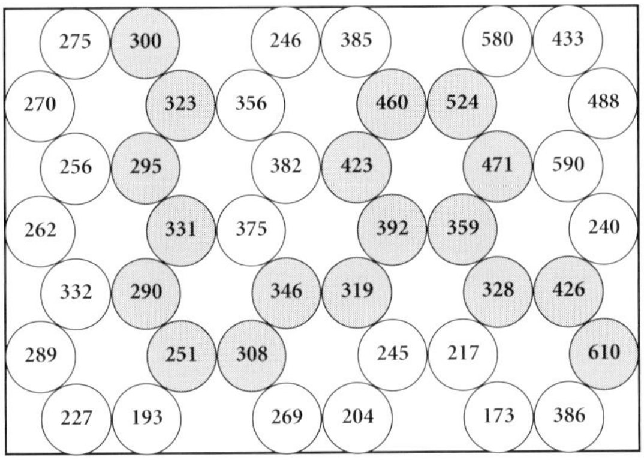

W4 Nearest hundred or thousand

1 (a) 400 (b) 700 (c) 1700
(d) 4600 (e) 12 600 (f) 18 500

2 (a) 43 000 (b) 19 000 (c) 3000
(d) 12 000 (e) 23 000

W5 Mileometer

1 100 miles

2 (a) 5000 miles (b) 70 miles
(c) 20 000 miles (d) 5000 miles

3 (a) 10 947 (b) 83 744 (c) 54 480

W7 Pinball

1 365

2 35 + 60 + 70 = 165

3 (a) 125 + 70 = 195
(b) 60 + 70 = 130
(c) 35 + 60 + 80 = 175
(d) 250 + 125 = 375
(e) 250 + 125 + 80 = 455
(f) 250 + 125 + 70 = 445
(g) 250 + 80 + 35 = 365
(h) 250 + 125 + 35 + 80 = 490
(i) 70 + 60 + 250 + 35 = 415
(j) 250 + 125 + 70 + 35 = 480

W9 Multiplication magic

12 345 679 × 4 = 49 382 716
and 49 382 716 × 9 = 444 444 444

12 345 679 × 5 = 61 728 395
and 61 728 395 × 9 = 555 555 555

12 345 679 × 6 = 74 074 074
and 74 074 074 × 9 = 666 666 666

and so on

W10 Remainder code

1 P 2 L

3 O 4 V

5 L 6 E

7 R 8 I

9 O LIVERPOOL

W11 Fruity sums

1 ORANGE 2 MELON

3 MANGO 4 LEMON

5 GRAPE 6 PRUNE

7 PEAR 8 APPLE

9 PLUM

W12 School roll

1 224 pupils

2 28 pupils in each class

3 629 girls, 596 boys, so 33 more girls

4 25 pupils

5 £1225 × 35 = £42 875

W13 Two sum puzzle

$$\begin{array}{r} 408 \\ +\underline{273} \\ \underline{681} \end{array} \qquad \begin{array}{r} 135 \\ +\underline{259} \\ \underline{394} \end{array}$$

Fractions (p 8)

F1 Transport

1	CAR	2	BIKE
3	HORSE	4	BARGE
5	SKIS	6	CAMEL
7	GLIDER	8	SKATES
9	DONKEY	10	BALLOON

F2 What fraction of …

1	$\frac{2}{3}$	2	$\frac{3}{4}$
3	$\frac{3}{5}$	4	$\frac{3}{8}$
5	$\frac{3}{9} = \frac{1}{3}$	6	$\frac{5}{7}$

F3 Fraction maze

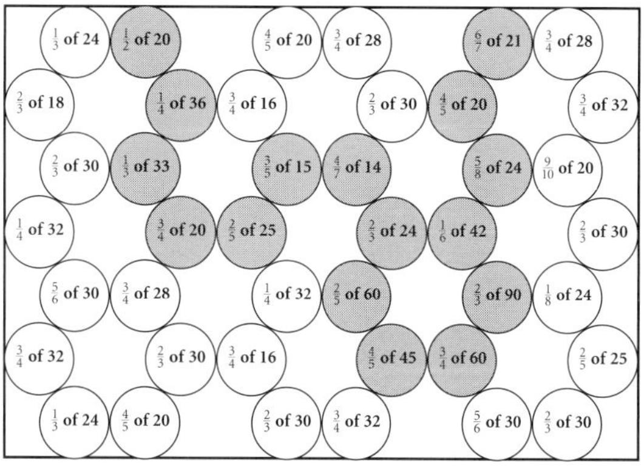

Decimals (p 9)

D1 Which fits?

1 3.9 cm by 2.4 cm

2 2.6 cm by 4.1 cm

3 2.5 cm by 3.8 cm

4 2.7 cm by 3.5 cm

5 2.6 cm by 3.7 cm

The hole measures 2.5 cm by 3.8 cm.
Rectangle 3 will fit into the white hole.

D2 Reading scales

1 (a) 0.6 (b) 1.3 (c) 2.1
(d) 2.7 (e) 3.1

2 (a) 5.8 (b) 7.5 (c) 8.5
(d) 10.1 (e) 6.5 (f) 7.5

D4 Jumping contest

1 Barry (3.9 m) 2 Derek (4.6 m)

3 Neil (5 m) 4 0.4 m (5 – 4.6)

D5 More or less

3.5 is 0.4 more than 3.1
2.9 is 2 less than 4.9
2.9 is 0.2 less than 3.1
5.7 is 0.5 less than 6.2
5.7 is 0.8 more than 4.9
6 is 0.2 less than 6.2
6 is 0.8 less than 6.8
6 is 0.9 less than 6.9
7.3 is 0.5 more than 6.8
7.3 is 0.4 more than 6.9

D7 Guessing games

1 (a) 16.3 m (b) 1.6 m

2 (a) 152 litres (b) 3.3 litres

D8 Addition crosses

1

	0.4	
1.6	0.7	**2.3**
	1.1	

2

	1.7	
2.3	0.8	**3.1**
	2.5	

3

	1.3	
2.5	4	**6.5**
	5.3	

4

	1.9	
1.4	**0.6**	2
	2.5	

5

	0.8	
2.1	1.3	3.4
	2.1	

6

	3.5	
4.8	**1.2**	6
	4.7	

7

	1.1	
0.3	1.7	**2**
	2.8	

8

	4.8	
1.6	0.9	2.5
	5.7	

D9 Make them right

1 0.2 **2** 3

3 5 **4** 5

5 0.6 **6** 0.2

D10 Packing a case

The best solution is brooch, computer, video and painting, having a total weight of 13.6 kg and value £1560.

D11 Shopping around

1 66p

2 (a) (i) Megavalue
(ii) Foodfair
(b) 21p

3 (a) £1.65 (b) £4.20

4 Costs are
Trusty £2.52
Sunbury's £2.54
Foodfair £2.10
Megavalue £2.29
So Foodfair is cheapest

2 Using a spreadsheet

This unit collects together a number of computer-based activities. They need not be done by pupils in one continuous period. Depending on a school's arrangements for computer access, these activities can be used individually in a computer period or be dipped into when the work is appropriate

Our expectation is that your pupils will already have met spreadsheets, though they may need reminding about how to use them.

Some activities may be done using a graphic calculator with easy-to-use spreadsheet facilities.

All of these activities will benefit from pupils working in pairs so that they can discuss their work.

Essential

Spreadsheet program

Spot the formula (p 14)

This activity can be used to remind pupils of *Book 1* unit 15, 'Spot the rule'.

◊ Some pupils may invent formulas that are impossibly hard to spot. If the game is to be fun for both players, you will need to discourage this.

◊ Opportunities for discussing equivalent formulas will undoubtedly arise. One pupil may hide a formula such as `= 2*A1 + 5`, and the second make a (correct) guess such as `= A1 + A1 + 5`. Discussion between pupils, and between yourself and pupils, of such equivalences will be invaluable.

◊ It will be interesting to discuss strategies for finding formulas with pupils. For example, putting 100 or 1000 into a formula can often tell you a lot.

Making a sequence (p 14)

This activity builds on *Book 1* unit 12, 'Work to rule', and links with unit 9 'Hexathings', in this book.

T

◊ Recent versions of spreadsheets such as Excel have shortcuts for filling a column with the numbers 1, 2, 3, ...
It is true that pupils could use a formula to put the numbers 1, 2, 3, … into column A. However, this draws attention to formulas that work down column B, and away from formulas that work across, connecting each number in column A and the corresponding number in column B.

◊ You may wish to discuss how to go about example 1 with the whole group. Even if it is done in pairs, it will be valuable to discuss with the group what formulas they have spotted. For example:

1 Put = `A1*2` into cell B1, and fill down column B.

2 Type 2 into cell B1. Put = `B1+2` into cell B2 and fill down.

Method 1 exemplifies spotting the *n*th term of the sequence. Method 2 parallels finding the term-to-term rule. Both methods are perfectly valid, and pupils can be encouraged to find yet others.

Solutions

1	`=2*A1`	2	`=A1+5`	3	`=3*A1`
4	`=5*A1`	5	`=5*A1+1`	6	`=5*A1–1`
7	`=2*A1+1`	8	`=10*A1–1`	9	`=A1*A1`

or equivalent formulas

Big, bigger, biggest (p 16)

◊ For some pupils it will not be immediately obvious that there may be decimal solutions to the problems. It will probably be best if pupils discover this for themselves. The second problem focuses attention on this, even if pupils have overlooked decimals when doing the first problem.

◊ The problems may be easier to solve if the number of variables is minimised. So in example 2, if two of the numbers are in cells B1 and B2, then into cell B3 put the formula `11–B1–B2` for the third number.

Solutions

1 (a) 6, 2
(b) 12, 2
(c) 11, 3; 10, 4; 9, 5; 8, 6; 7, 7 (plus an infinite number of decimal solutions) all give numbers greater than 24 when multiplied. The biggest number when multiplied is 7 × 7 = 49.

2 (a) 1, 4, 6 make 24.
(b) The biggest number is 3.6 × 3.6 × 3.6 (or 3.7 × 3.7 × 3.7). Pupils may need nudging to find a solution using decimals.

Furry Festivals (p 17)

T

In this activity pupils should be encouraged to use methods that restrict the number of variables. For example, in the second problem they might make the number of pens equal to 60 minus the numbers of T-shirts and badges.

- ◊ The spreadsheet may be easier to understand if the cells containing prices are formatted to show a £ sign.
- ◊ The problem can be extended by including another item to be sold, or prices that are not whole numbers.

Solutions

1. At the first festival they sold 20 T-shirts and 20 badges.
   At the second festival they sold 28 T-shirts and 12 badges.
2. 5 T-shirts, 35 badges and 20 pens, or 6 T-shirts, 14 badges and 40 pens are both solutions. At first sight 4 T-shirts, 56 badges and 0 pens also works, but we are told that they sold at least one of each item.

3 Using rules

Practice booklet pages 3 to 5

A Katy's Catering (p 18)

◊ The teacher-led introduction can be used to discuss with pupils how many of each item would be needed for a party for different numbers of people. Any even number of people gives whole-number answers but 10 and 20 are used in section A questions so should be avoided.

You could also at this stage talk about how the rules can be shortened into the form used in section B.

◊ Pupils may need help in understanding how to translate the wording of each of the formulas on the first page into mathematical operations.

◊ Pupils with reading difficulties may find this section rather difficult and some schools have preferred to do the whole section orally.

B Shorter rules (p 19)

'A lot easier once the rules were shortened'

◊ You may wish to discuss how to shorten the rules with the whole class if you did not do so at the start of section A.

B3 You may need to reinforce the 'fraction' notation for division.

C Shorthand (p 20)

This section introduces letters standing for numbers. It is very important to emphasise that n stands for 'the **number** of people' and not for the word 'people'. It is a good idea sometimes to choose a letter that is different to the first letter of the object referred to.

◊ The teacher-led introduction can be used to discuss with pupils how to write the word formulas in the normal mathematical shorthand. You can use the formulas in section B as a basis for this, or get the pupils to suggest formulas of their own for writing in shorthand.

Include discussion of substituting into the shorthand formulas, so that pupils can see how they work.

◊ Use the opportunity to discuss such common mistakes as:

- Pupils writing $p \times 2$ as $p2$ and not (as convention demands) $2p$.
- Substituting (say) $p = 3$ into $2p$ and getting 23.
- Not realising that p and $1p$ are the same.

◊ You may need to explain the meaning of
'If $m = p + 6$, what is m when $p = 3$?'

◊ Much of this unit lends itself to treatment on a spreadsheet. For example:

	A	B	
1	**Katy's Catering**		
2	Number of people	10	
3			
4	Chicken legs	=B2+6	
5	Paper cups	=B2*2	
6	Lemonade	=B2/2	
7	Sausage rolls	=3*B2+10	

'We decided to use a spreadsheet. When we changed the numbers – WOW – how quickly all the answers change!'

Getting the pupils themselves thinking about what formulas are appropriate can greatly help their understanding of the algebra involved.

D Working out formulas (p 23)

D3 Watch out for the rule $2p + 50$, where the pupil has simply translated '2 pencils and 50 more' into pseudo-mathematics.

A Katy's Catering (p 18)

A1 (a) 16 (b) 26

A2 (a) 20 (b) 40

A3 (a) 5 (b) 10

A4 (a) 40 (b) 70

A5 15

A6 (a) 320 (b) 620

B Shorter rules (p 19)

B1 (a) 20 (b) 32

B2 (a) 14 (b) 24

B3 (a) 10 (b) 8 (c) 50

B4 (a) 5 (b) 15

C Shorthand (p 20)

C1 (a) 18 (b) 14 (c) 26 (d) 38

C2 (a) 24 (b) 36 (c) 30 (d) 60

C3 (a) 160 (b) 45 (c) 8

C4 (a) 80 (b) 400 (c) 20
(d) 20 (e) 12

C5 (a) 70 (b) 55 (c) 160

C6 (a) 55 (b) 115 (c) 5, 15

C7 $c = n + 6$

C8 $s = 3n$

*C9 $l = \frac{n}{2}$

*C10 $r = 2n + 6$

D Working out formulas (p 23)

D1 (a) (i) 5 (ii) 242 (iii) 78 (iv) 26
(b) 3 tents, 146 packets of porridge, 46 sardine cans, 18 ropes
(c) 10 tents, 482 packets of porridge, 158 sardine cans, 46 ropes

D2 (a) 50 textbooks, 180 mirrors, 850 cubes, 10 wallcharts
(b) 25 textbooks, 80 mirrors, 450 cubes, no wallcharts
(c) 250 textbooks, 980 mirrors, 4050 cubes, 90 wallcharts

D3 If p is the number of pencils she buys, then $p = 2c + 50$.

D4 (a) 24 chairs, 46 bibs, 90 blocks, 11 tables, 38 tidyboxes
(b) 10 chairs, 18 bibs, 20 blocks, 4 tables, 10 tidyboxes
(c) 14 chairs, 26 bibs, 40 blocks, 6 tables, 18 tidyboxes

D5 (a) 20 (b) 35 (c) 8 (d) 5

D6 (a) 34 (b) 4 (c) 0 (d) 194

D7 (a) 4 (b) 19 (c) $6\frac{1}{2}$ (d) 44

*D8 (a) 40 (b) 25 (c) 60

*D9 (a) 8 (b) 8 (c) 180

What progress have you made? (p 25)

1 54 2 11 3 5

4 20 5 13

6 (a) 43 (b) 11 (c) 403

7 (a) 11 (b) 39 (c) 399

Practice booklet

Sections A and B (p 3)

1 (a) 14 desks (b) 20 children
(c) 20 children (d) 29 worksheets
(e) 30 children

2 (a) 7 tables (b) 32 children
(c) 24 children (d) 120 pencils

Section C (p 4)

1 (a) 5 bags (b) 60 bushes
(c) $s = 16$ (d) $c = 16$
(e) You need **4** trays of petunias for each flower bed and another **3** extra.

2 $w = 4b$

3 $g = 5b + 4$

Section D (p 5)

1 (a) 2 sheets of plywood, 10 hinges, 90 screws, 130 nails, 46 tacks, 20 labels
(b) 5 sheets of plywood, 25 hinges, 210 screws, 310 nails, 106 tacks, 50 labels

2 (a) 6 cans (b) 4.5 cans or $4\frac{1}{2}$ cans

3 (a) 14 (b) 32 (c) 2 (d) 242

4 (a) 2 (b) 1 (c) 6 (d) 10

5 (a) 41 (b) 1 (c) 96 (d) 496

6 (a) 10 (b) 15 (c) 30 (d) 255

4 Repeats (p 26)

The main purpose of this unit is revision of coordinates.
It also introduces translations in an informal way.

T

◊ The introductory activity might go as follows.

- Ask simple questions to revise coordinates such as 'What are the coordinates of the nose on the first ghost?'
- Ask what change is made to the coordinates when moving from one nose to another along a row.
  What change occurs when moving to the nose on the row above?
- Check that the same rules apply to other parts of the ghost.
- Calculate the cooordinates of shapes beyond the grid.

4(b) This involves using negative coordinates.

1 (a) (2, 2)
(b) Add 2 to the x-coordinate.
(c) (i) (8, 2) (ii) (10, 2)
(d) Add 3 to the y-coordinate.
(e) (i) (2, 8) (ii) (6, 8)
(f) (i) (10, 5) (ii) (8, 8)
(iii) (10, 8)

2 (a) A (3, 2) B (6, 3)
(b) (i) Add 3 to the x-coordinate.
(ii) Add 1 to the y-coordinate.
(c) Yes
(d) (12, 5)

3 (a) You add 2 to the x-coordinate and add 4 to the y-coordinate.
(b) (11, 8) (c) (13, 12)

4 (a) (12, 9) (b) $(2, {}^{-}1)$

5 Multiples and factors

Essential
Sheets 134, 220
Sheet 176 or 221 (221 is easier)
Coloured pens or pencils
Dice, counters (two colours)

Practice booklet pages 6 to 7

Optional
OHP transparency of sheet 134

A Multiples (p 28)

'The pupils love colouring in and quickly recognised the patterns formed.'

Sheet 134, coloured pens or pencils
Optional: OHP transparency of sheet 134

You may wish to introduce this activity using an OHP. Pupils need to be aware that multiples of 3 (for example) continue beyond the end of the 3 times table.

B Multiplying (p 28)

The activities described in unit 5, 'Multiplication', in *Book N* can also be used (for example, 'Tables bingo').

C Missing multiples (p 30)

'The "Nasty multiples game" is very good. I enjoyed playing it and the pupils sometimes beat me.'

Sheet 220
One dice and several counters of two colours for each pair.

D Factors (p 31)

T

The factor spider helps pupils to be systematic when looking for factors of a number. When making a factor spider, start with 1 and write the factors in order round the spider.

Factors bingo

This is a useful and fun way to practise finding factors.

◊ Ask each pupil to write down seven numbers which are to be factors. Numbers can be repeated (for example, 2, 2, 3, 3, 4, 6, 7). They are not allowed to use 1, but if they suggest it, reward their perspicacity!

Make a list of 'multiples' for yourself and read them out one at a time. Include the occasional prime number!

Pupils cross out all the factors of the number you read out. They are not allowed to cross out a number more than once. For example, if a pupil's list includes 2, 2, 4, 8 and the number read out is 16, they can cross out **one** of the 2s, the 4 and the 8.

For the first couple of games you can check after each number what factors could have been crossed off.

◊ An alternative approach is for the pupils to have, or make, number cards which they hold up. If they hold up a correct card they can throw it out; if a card is incorrect you can give them an extra card as a penalty.

E Factor codes (p 32)

'A brilliant exercise – they loved it!'

Sheet 176 or 221. (Sheet 221 is easier.)

When pupils try to decode messages, they may find it helpful to first use a factor spider to write down all the factors of the number of letters.

Solutions to sheets 176 and 221 are included in the answer section.

B Multiplying (p 28)

B1 5 × 4 = **20**

B2 (a) 3 × 4 = 12 (b) 2 × 8 = 16
(c) 6 × 3 = 18

B3 (a) 3 × 5 = 15 (b) 2 × 6 = 12
(c) 5 × 5 = 25

B4 (a) 14 (b) 15 (c) 10
(d) 30 (e) 35

B5 (a) 2 × **5** = 10 (b) 3 × **3** = 9
(c) **8** × 2 = 16 (d) 5 × **6** = 30
(e) 4 × **3** = 12

B6 (a) Yes (b) No (c) No (d) Yes

B7 (a) No (b) No (c) Yes (d) No

B8 3, 6, 9, 12, 15, 21, 30

B9 5, 10, 15, 20, 25, 45, 50, 55

B10 (a) 6, 10, 14, 18, 20, 26, 30
(b) Even numbers

B11 (a) The pupil's five multiples of 10
(b) 70, 90, 100, 110, 220
(c) Its last digit is 0.

B12 (a) 20, 22, 24, 26, 28, 30
(b) 21, 24, 27, 30
(c) 20, 24, 28
(d) 20, 25, 30 (e) 24, 30
(f) 21, 28 (g) 24
(h) 27 (i) 20, 30 (j) 22

C Missing multiples (p 30)

C1 (a) 48 (b) 35 (c) 54

C2 8

C3 (a) 5 (b) 7 (c) 12

C4 (a) 4 × 8 = 32 (b) 3 × 9 = 27
(c) 7 × 5 = 35 (d) 6 × 6 = 36
(e) 7 × 4 = 28

C5 (a) 6 × **4** = 24 (b) 4 × **9** = 36
(c) **9** × 5 = 45 (e) **3** × 7 = 21

C6 (a) 6 × **5** = 30 (b) 7 × **6** = 42
(c) **9** × 6 = 54 (d) **4** × 7 = 28

C7 1 × 12 = 12, 2 × 6 = 12, 3 × 4 = 12
(and their 'reverses')

C8 1 × 20 = 20, 2 × 10 = 20, 4 × 5 = 20
(and their 'reverses')

C9 (a) 6, 8, 12, 18, 20, 24, 30
(b) 6, 12, 15, 18, 24, 30
(c) 8, 12, 20, 24
(d) 15, 20, 30

C10 33p, 59p, 66p

C11 1 × 30 = 30, 2 × 15 = 30, 3 × 10 = 30,
5 × 6 = 30 (and their 'reverses')

C12 (a) 6 is a multiple of 1, 2, 3 and 6.
(b) 8 is a multiple of 1, 2, 4 and 8.
(c) 12 is a multiple of 1, 2, 3, 4, 6 and 12.
(d) 20 is a multiple of 1, 2, 4, 5, 10 and 20.

D Factors (p 31)

D1 (a)
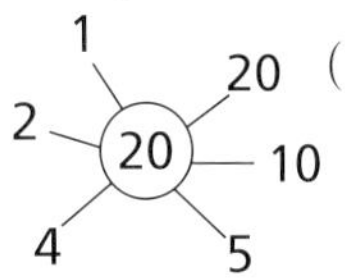

(b)
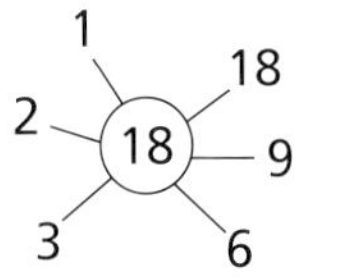

(c)
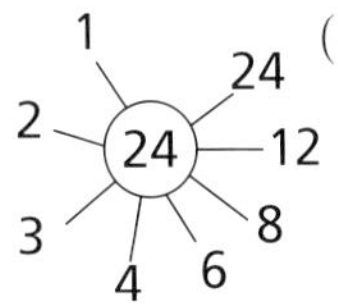

(d)
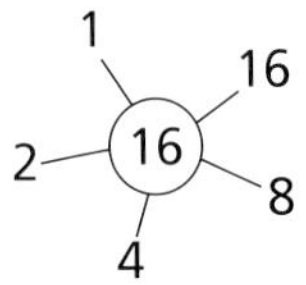

D2 6

D3 20, 40

D4 (a) 6 is a **multiple** of 2.
(b) 6 is a **factor** of 30.
(c) 10 is a **factor** of 60.
(d) 8 is a **factor** of 32.
(e) 8 is a **multiple** of 4.
(f) 36 is a **multiple** of 9.
(g) 15 is a **factor** of 30.
(h) 6 is a **multiple** of 6 and also a **factor** of 6.

E Factor codes (p 32)

E1 Party at ten (5 × 2 grid)

E2 (a) Meet me at four (4 × 3 grid)
(b) Train leaves at eight (3 × 6 grid)
(c) Leave bag behind bus shelter (3 × 8 grid)
(d) Will see you outside cinema tonight (5 × 6 grid)

E3 The pupil's message

Multiples and factor maze

Sheet 176

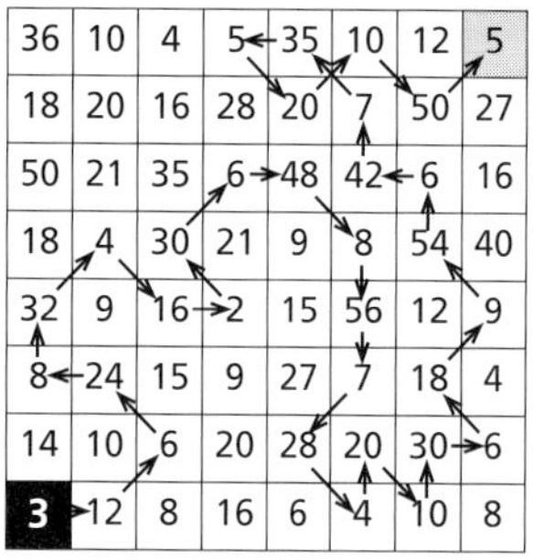

Sheet 221

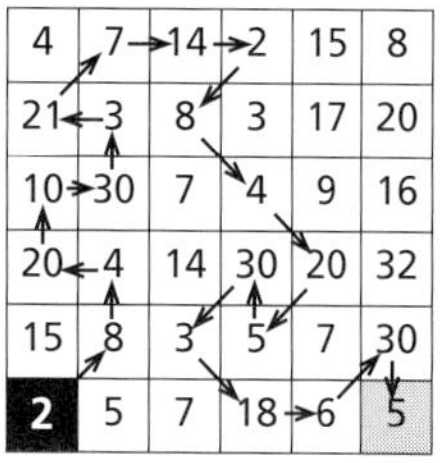

What progress have you made? (p 33)

1 The pupil's three multiples of 4

2 6, 12, 18, 24

3 1, 2, 4, 5, 10, 20

4 1, 2, 3, 5, 6, 10, 15, 30

5 (a) 15 is a **multiple** of 5
(b) 4 is a **factor** of 32
(c) 12 is a **factor** of 24
(d) 36 is a **multiple** of 12

Practice booklet

Sections B and C (p 6)

1 3 × 5 = **15**

2 (a) 4 × 3 = 12 (b) 3 × 6 = 18
(c) 6 × 4 = 24 (d) 6 × 5 = 30

3 (a) 16 (b) 35 (c) 18
(d) 48 (e) 32

4 (a) 5 × **2** = 10 (b) 4 × **3** = 12
(c) 6 × **5** = 30 (d) **12** × 2 = 24
(e) **10** × 3 = 30 (f) **12** × 5 = 60

5 (a) 4, 6, 8, 10, 20, 50
(b) 6, 9 (c) 5, 10, 20, 35, 50
(d) 10, 20, 50

Sections D and E (p 7)

1 (a) factor (b) multiple
(c) multiple (d) multiple
(e) multiple and factor
(f) multiple (g) factor
(h) factor (i) multiple

2 1, 2, 3, 5, 6, 10, 15, 30

3 6 and 36

4 (a) PEAR (b) PLUM (c) GRAPE

5 (a) Smile at your teacher
(b) You have finished your homework

6 Puzzle it out

T	p 34 **A** Road map	The map is a basis for asking oral questions
T	p 34 **B** Count up	A game giving number practice

Essential	**Optional**
Large quantities of rough paper	Numbered cards (see section B)
Practice booklet page 8	

A Road map (p 34)

T

'I used this as practice for SATs, reading each question out twice.'

◊ The map is intended to be used as a source for your own oral questions. It shows distances and journey times in East Anglia.

◊ At first it is better to ask your questions in two stages.
First ask the pupils to find the road or roads, then ask the question.
For example:

'Find the road that goes from Swaffham to Thetford and then on to Bury St Edmunds.
How long does it take to go from Swaffham through Thetford to Bury St Edmunds?'

◊ Some different types of question are given below. These are just examples.

	Basic questions: reading the map	**Answer**
1	How far is it from Swaffham to Norwich?	28 miles
2	If you go from Swaffham to Bury St Edmunds, which town do you go through?	Thetford
3	How long does it take to go from Great Yarmouth to Cromer?	50 min
4	Which town is 27 miles from Norwich?	Fakenham
5	How far is it from Ipswich to Great Yarmouth?	54 miles
	Adding distances	
1	How far is it from Swaffham through Thetford to Bury St Edmunds?	32 miles
2	How far is it from Swaffham to Cromer through Fakenham?	38 miles
3	How far is it from Thetford to Swaffham and back again?	40 miles
4	How far is it from Ipswich to Cromer through Great Yarmouth?	87 miles
5	How far is it from Ipswich to Cromer through Norwich?	66 miles

Adding and subtracting times: answers in minutes

1 How many minutes does it take to go from Norwich to Bury St Edmunds through Thetford? — 55 min

2 How many minutes does it take to go from Fakenham to Great Yarmouth through Norwich? — 75 min

3 Find the road from Norwich to Fakenham and the road from Norwich to Great Yarmouth.
How much longer does it take to get to Fakenham than to Great Yarmouth? — 15 min

4 How much longer does the journey from Norwich to Ipswich take than the journey from Norwich to Thetford? — 25 min

5 You want to go from Fakenham to Great Yarmouth. You could go through Cromer or through Norwich.
Which way is quicker? — Norwich
How much quicker is it? — 15 min

Subtracting distances

1 Find Fakenham. Which is further from Fakenham, Swaffham or Cromer? — Cromer
How much further is it? — 6 miles

2 Find Bury St Edmunds. How much further is it to Ipswich than to Thetford from Bury St Edmunds? — 15 miles

3 From Ipswich it is further to Great Yarmouth than to Norwich.
How much further is it? — 11 miles

4 From Norwich it is further to Ipswich than to Thetford.
How much further is it? — 14 miles

5 How much further from Norwich is Fakenham than Great Yarmouth? — 8 miles

Working out arrival and departure times

1 You are going from Ipswich to Norwich. You leave Ipswich at 11:15.
What time do you get to Norwich? — 12:15

2 You are going from Swaffham to Norwich. You leave Swaffham at 9:30.
What time do you get to Norwich? — 10:10

3 You are going from Norwich to Cromer. You leave Norwich at 4:45.
What time do you get to Cromer? — 5:20

4 You are going from Swaffham to Thetford.
You need to be in Thetford by 5:30.
What is the latest time you can leave Swaffham? — 4:50

5 You are going from Great Yarmouth to Ipswich.
You leave Great Yarmouth at 3:45.
What time do you get to Ipswich? — 5:05

B Count up (p 34)

T

'A really nice activity which lots of staff have tried.'

> Large quantities of rough paper
> Optional:
> Four sets of numbered cards for the teacher to use:
> set A – 1, 1, 2, 2, 3, 3, 4, 4, 5, 5
> set B – 6, 6, 7, 7, 8, 8, 9, 9, 10, 10
> set C – 1, 1, 2, 2, 3, 3, 4, 4, 5, 5, 6, 6, 7, 7, 8, 8, 9, 9
> set D – 10, 10, 20, 20, 25, 25, 50, 50

◊ The aim is to reach a target number using a set of six numbers. Each number can only be used the number of times that it appears. There may be several ways of reaching a target or there may be none (in which case the aim is to get as close to the target as possible).

For example, with 4, 1, 5, 4, 7, 10 you can reach the target 73 like this:

$$\begin{array}{r} 7 \times 10 = 70 \\ +\ 4 \\ -\ 1 \\ \hline 73 \end{array}$$

Playing the game

With the cards While the game can be played without cards, pupils enjoy it more if you use the sets of cards detailed above. Put each set of cards face down. Then select cards from the sets as follows.

'Using the cards is really good. I used a graphical calculator to get a random target number.'

Game 1

- Choose four cards from set A and two from set B.
- Choose (or randomly generate) a target roughly in the range 10 to 100.

Game 2

- Choose four cards from set C and two from set D.
- Use a target roughly in the range 100 to 200.

Get pupils to write down the numbers on the chosen cards. After the target number has been selected, pupils start working on reaching it.

You can introduce a time element (for example, everyone stops after one minute); alternatively, everyone stops when the first pupil announces that they have made the target number.

Ask a pupil to describe how they got their number. Write the calculations up as they are explained so that any errors can be spotted and discussed. Alternative methods of reaching the target can be discussed.

Without the cards Give the pupils the six numbers, which you might generate randomly or just choose yourself. The target number can be chosen by a pupil, by yourself or generated randomly. Vary the target number depending on the degree of difficulty required. Numbers less than 100 are easiest, between 100 and 200 slightly harder, and numbers greater than 200 often cause problems.

◊ The idea is to encourage quick and accurate mental arithmetic, but pencil and paper methods can be used. Pupils should not be restricted by concerns about the presentation of their work, hence large quantities of rough paper should be available.

Don't require pupils to use brackets in their solutions, but you might afterwards discuss how to write out solutions 'properly'.

◊ Here is one way of achieving the best possible result for each of the examples on page 34. There may be others.

Target 73: $7 \times 10 = 70$, $70 + 4 - 1 = 73$

Target 123: $3 + 9 = 12$, $2 + 8 = 10$,
$10 \times 12 = 120$,
$120 + 5 - 2 = 123$

Target 169: The nearest is 168 or 170.
168 $25 - 10 = 15$, $15 \times 5 \times 2 = 150$,
$150 + 3 \times 6 = 168$
170 $2 + 6 = 8$, $25 - 8 = 17$,
$17 \times 10 = 170$

Practice booklet

Section B (p 8)

Small numbers

1 $5 \times 2 \times 8$

2 $(4 + 3) \times 7 - 2$

3 $6 \times 8 - 3$

4 $(3 + 4) \times (8 + 2)$

5 $4 \times 5 \times 5 + 2$

6 $2 \times 5 \times 8 - 3$

7 $2 \times 5 \times (6 - 1) + 2$

8 $(3 + 3 + 4) \times 2 - (1 + 2)$

9 $(2 + 3) \times 2 + 1 + 1 + 2$

10 $(2 + 3 + 5) \times 6 + 1 + 4$

Larger numbers

11 $10 \times 25 - (2 + 3) \times 9$

12 $8 \times 25 + 50 + 4 - 3$

13 $(6 + 2) \times 50 + 4 \times 7$

14 $(5 + 7) \times 25 + 25 + 8 - 3$

15 $50 \times 2 + 5 \times 10 - (1 + 2) = 3$

16 $10 \times 50 - 5 \times 8 + 2 \times 3$

17 $3 \times 3 \times 50 + 25 - 6 - 7$

18 $(2 + 8) \times 10 \times 10 + 1$

19 $50 \times 50 - 4 \times 9$

20 $((50 - 9) \times 10 - 6) \times (2 + 3)$

7 Two decimal places

This unit extends decimal notation to two decimal places.

Essential	**Optional**
Sheets 223 and 224 Metre rules	Tape measure marked in decimals of a metre OHP transparency of sheet 222 Large cards with digits 0 to 9 on them
Practice booklet pages 9 to 13	

A Tenths review (p 35)

B Tenths and hundredths (p 36)

Optional: OHP transparency of sheet 222

◊ The diagram on page 36 shows a litre measuring cylinder marked in tenths and hundredths. You can ask why the new subdivisions are called 'hundredths', and relate this to the fact that there are 100 of them in the whole unit.

◊ It is useful to have one of these 'measuring cylinders' copied on to an OHT, using sheet 222. By cutting out the shaded strip, or using a coloured piece of acetate, you can then vary the height of the shaded column and ask for the measurement in decimal form.

◊ Emphasise the place values: tenths, hundredths.

◊ Numbers below 0.1 are dealt with in section C, but you may wish to introduce them here.

C Less than 0.1 (p 38)

Sheet 223 for question C3
Optional: OHP transparency of sheet 222

T

◊ The OHT of sheet 222 is useful again here.

◊ Pupils often find it more difficult to recognise decimals like 0.02 when reading a scale. This may be because in other instances the tenths mark below shows how the decimal begins. (For example, 0.42 is above 0.4.)

D 0.3 and 0.30 are equal (p 39)

T

'I emphasised the usefulness of using decimals with the same number of decimal places, in particular for addition, subtraction and writing in order.'

This is another aspect of decimals that confuses pupils, made worse by their tendency to read, for example, 1.30 as 'one point thirty' (like 'one pound thirty').

Even pupils who will agree that 3 and 3.0 are the same are often less confident about 0.3 and 0.30.

E Decimals of a metre (p 40)

Sheet 224, metre rules
Optional: Tape measure marked in decimals of a metre

T

Practical activities

Most will be gained from this section by the use of practical measuring activities. These activities should include measurements where the tenths digit is 0, as in 2.08 m.

Long tape measures, as used by PE departments, often have centimetre scales which are marked in intervals of 0.1 m, for example 2.10, 2.20 etc. These may be useful in illustrating the idea of decimals of a metre.

Some suggested activities are:

Cubits
The ancient measurement of a cubit was said to be the distance between the elbow and the tip of the middle finger. Use blocks of wood or books on a desk at the elbow and fingertips to measure accurately the distance between them to the nearest hundredth of a metre. The class could discuss how accurate a measurement this was.

Standing high jump

First measure the 'reach' of pupils from the floor to their outstretched arms, in metres to two decimal places. Then ask pupils to stand in front of a wall holding a piece of chalk and jump to make a mark as high as they can on the wall. The heights jumped can also be recorded as decimals of a metre. If pupils work in small groups they can list their group's jumps in order of size. The difference between their arms' outstretched 'reach' and the jump height can be used as a fairer assessment of pupils' jumping abilities. A standing long jump event could also be staged.

Records

A tape measure could be laid out on a floor (or pinned to a wall). Results from a recent athletics long jump or historical records could be written on arrow-shaped cards and placed as appropriate on the line. For example some recent winning distances in the Olympic women's long jump were

1948	5.69 m	1976	6.72 m
1952	6.24 m	1980	7.06 m
1956	6.35 m	1984	6.96 m
1960	6.37 m	1988	7.40 m
1964	6.76 m	1992	7.14 m
1968	6.82 m	1996	7.12 m
1972	6.78 m		

Discussion points

◊ The three bulleted questions on the pupil's page raise crucial points for the understanding of decimals.

Height and armspan

◊ This gives further practice in measuring and recording. Measuring needs to be to two decimal places as the two measurements are close.

F Ordering decimals (p 42)

G Place value (p 43)

Optional: Large cards with the digits 0 to 9 on them

T

The activity below was used in *Book T1*, unit 5, 'Decimals 1', section D but can extended to two decimal places.

Give each pupil a card with a digit on it. Rule large 'columns' on the board labelled hundreds, tens, units, tenths and hundredths. Give the pupils a number, say 23.46. Invite pupils to come forward and hold the cards in the correct column. Then ask them to make a new number by adding say 0.01 to the number they have already. For example the pupil holding the 6 in the hundredths column would need to be replaced by a pupil holding a 7.

Alternatively this could be done by writing the numbers in columns on the board or OHP and asking pupils to carry out the additions by only changing **one** digit.

A Tenths review (p 35)

A1 (a) 2.3 (b) 1.4 (c) 0.6
(d) 3.9 (e) 3.4 (f) 4.5

A2 2.5

A3 3

A4 0.9

A5 0.9, 2.8, 3, 4.6, 5

A6 3.3

A7 (a) 3.7 (b) 8.2 (c) 7.6 (d) 7.6

A8 (a) 3.4 (b) 5.6 (c) 5.7 (d) 2.6

A9 (a) 24.6 (b) 33.6 (c) 23.7
(d) 155.6 (e) 145.7 (f) 9.9

A10 (a) 3 tens or 30 (b) 7 tenths or 0.7
(c) 4 tenths or 0.4

B Tenths and hundredths (p 36)

B1 (a) 0.26 (litre) (b) 0.53
(c) 0.63 (d) 0.88
(e) 0.11 (f) 0.97

B2 (a) 0.13 (b) 0.24 (c) 0.29
(d) 0.34 (e) 0.38 (f) 0.45

B3 (a) 0.25 (b) 0.15
(c) 0.35 (d) 0.65

B4 (a) 2.64 cm (b) 2.18

B5 (a) 6.23 (b) 6.37 (c) 6.42
(d) 6.47 (e) 6.51 (f) 6.58

B6 (a) 6.25 (b) 6.45
(c) 6.55 (d) 6.75

B7 (a) 4.51 (b) 4.58 (c) 4.65
(d) 4.72 (e) 4.8

C Less than 0.1 (p 38)

C1 (a) 0.04 (b) 0.07
(c) 0.02 (d) 0.05

C2 (a) 0.01 (b) 0.05
(c) 0.09 (d) 0.17
(e) 0.24 (f) 0.33
(g) 0.4 (h) 0.45

C3 (a)

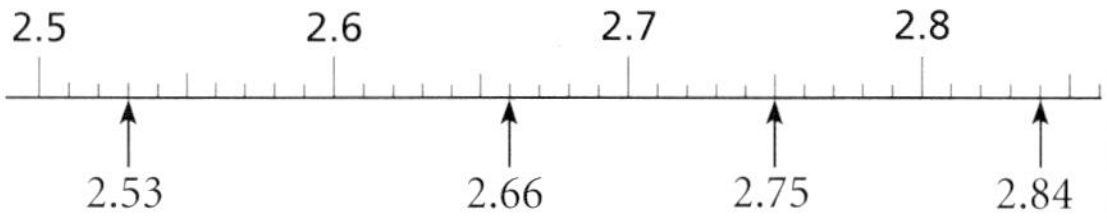

(b)

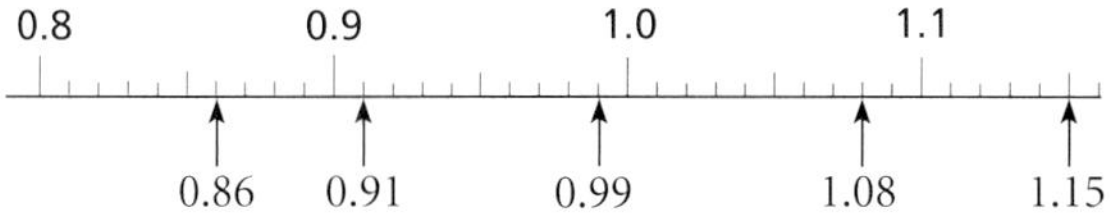

(c)

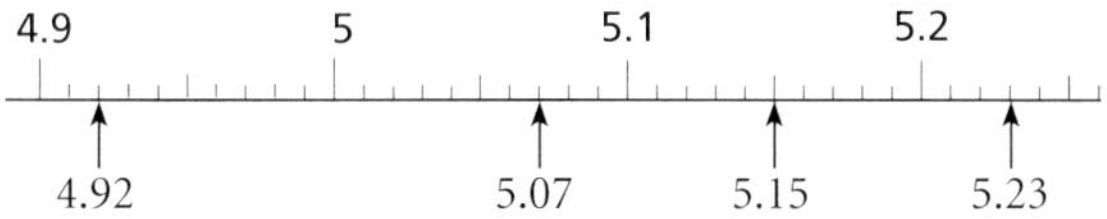

(d)

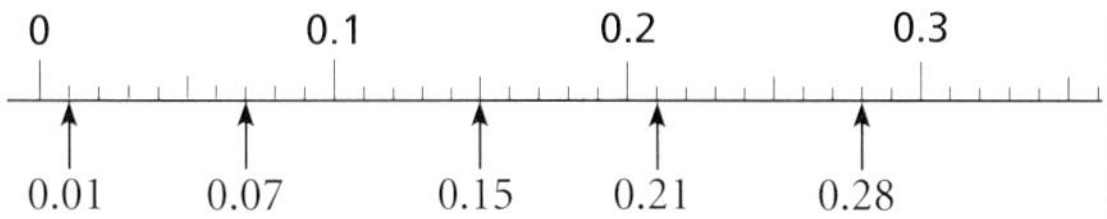

C4 (a) 0.05 (b) 0.09 (c) 0.01

D 0.3 and 0.30 are equal (p 39)

D1 (a) 0.70 (b) 0.10 (c) 0.8
(d) 0.4

D2 (a) 0.75 (b) 0.95 (c) 0.05

D3 0.05, 0.3 , 0.5

D4 (a) 0.2, 0.65, 0.9
(b) 0.08, 0.15, 0.4
(c) 0.07, 0.1, 0.3, 0.62

D5 (a) 0.03, 0.25, 0.6, 0.61
(b) 0.06, 0.1, 0.17, 0.3
(c) 0.08, 0.5, 0.55, 0.6

E Decimals of a metre (p 40)

E1 (a) 3.25 m (b) 4.68 m

E2 (a) 1 m 39 cm (b) 1.39 m

E3 (a) 1.7(0) m (b) 1.07 m
(c) 2.45 m (d) 3.06 m

E4 (a) 1.47 m (b) 1.07 m (c) 1.6(0) m

E5 The three rows of the table are:

Alan	167 cm	**1 m 67 cm**	**1.67 m**
Kira	**108 cm**	1 m 8 cm	**1.08 m**
Greg	**140 cm**	**1 m 40 cm**	1.4 m

E6 (a) Cobra (3.05 m)
(b) Boa constrictor (3.3 m)

E7 1.18 m, 1.4 m, 1.66 m

E8 0.09 m, 0.53 m, 0.7 m

E9 (a) 4.62 m (b) 4.65 m
(c) 4.68 m (d) 4.73 m

E10 (a) 7.01 m (b) 7.03 m
(c) 7.07 m (d) 7.11 m

E11 (a)

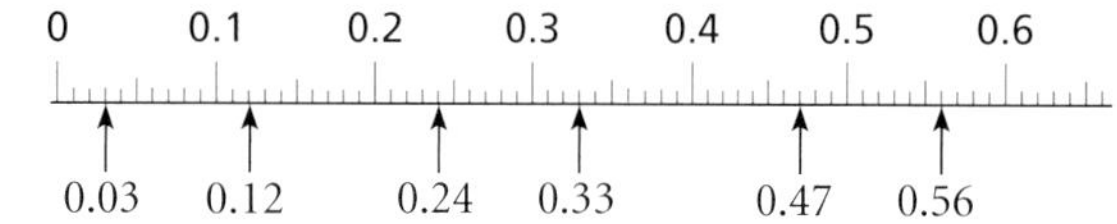

(b)

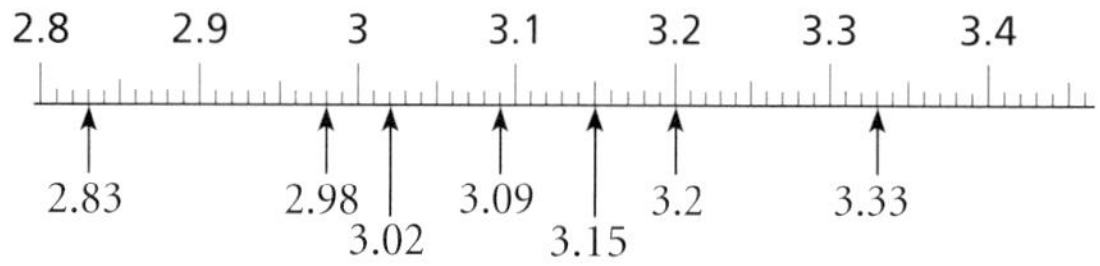

(c)

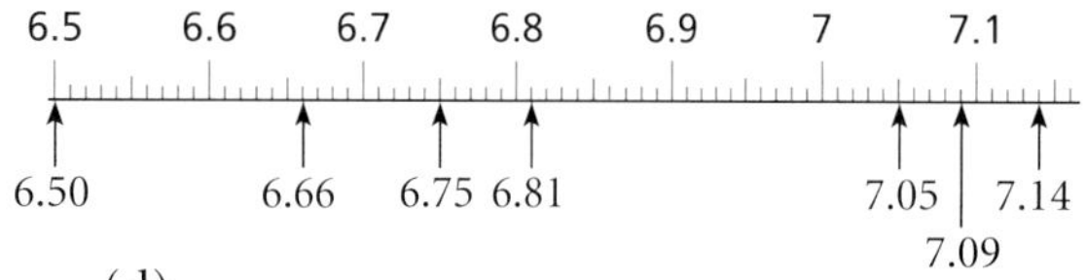

(d)

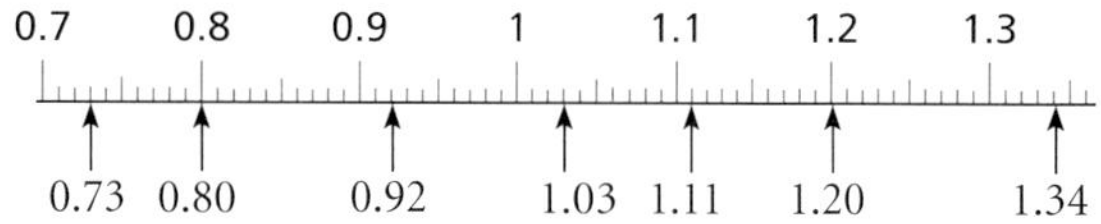

F Ordering decimals (p 42)

F1 2.35, 2.4

F2 5.17, 5.03, 5.1

F3 (a) 5.09, 5.32, 5.76, 5.84, 6
(b) 2.8, 3.07, 3.19, 3.2, 3.5
(c) 0.15, 0.45, 0.6, 1.07, 1.1

F4 HOLIDAYS

F5 ENJOYABLE

F6 DINOSAUR

F7 (a) 4.2 m (b) Barker
(c) Church (d) 4.5 m

G Place value (p 43)

G1 (a) 7 hundreds or 700
(b) 2 thousands or 2000
(c) 3 units or 3
(d) 1 ten or 10
(e) 8 hundredths or 0.08

G2 (a) 2723.58 (b) 3713.58
(c) 2713.68 (d) 2714.58
(e) 2713.59 (f) 2813.58

G3 (a) 1526.35 (b) 1427.35
(c) 1426.45 (d) 2426.35
(e) 1436.35 (f) 1426.36

G4 (a) 0.7 (b) 0.03 (c) 60
(d) 0.8 (e) 0.04 (f) 0.5

G5 60

G6 0.7

G7 0.03

What progress have you made? (p 44)

1 (a) 4.84 (b) 4.91 (c) 5.02 (d) 5.08
(e) 0.06 (f) 0.14 (g) 0.19 (h) 0.28

2 (a) 1.45 m (b) 2.03 m
(c) 3 m 20 cm

3 (a) 6.68, 6.7, 6.9, 7.04, 7.3
(b) 0.07, 0.1, 1.56, 3.24, 4

4 (a) 8 tenths or 0.8
(b) 4 hundredths or 0.04
(c) 7 hundredths or 0.07

Practice booklet

Sections B and C (p 9)

1 (a) 0.34 (b) 1.26 (c) 0.98
(d) 0.03 (e) 3.51

2 (a) 0.14 (b) 0.23 (c) 0.28
(d) 0.32 (e) 0.37 (f) 0.43

3 (a) 3.94 (b) 3.99 (c) 4.02
(d) 4.07 (e) 4.15 (f) 4.22
(g) 4.26 (h) 4.31

4 (a) 4.15 (b) 4.25
(c) 4.05 (d) 3.95

5

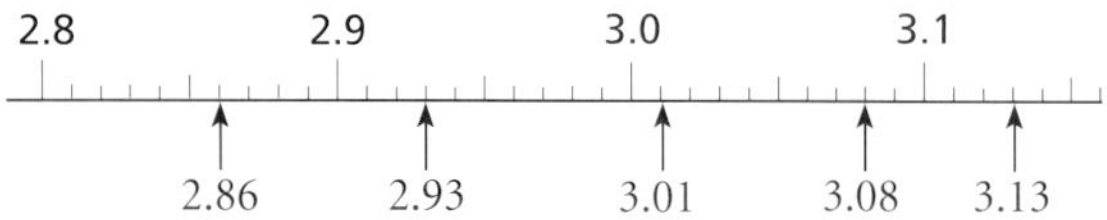

Section D (p 10)

1 (a) 0.55 (b) 0.25 (c) 0.05 (d) 0.45

2 A and H (0.7 and 0.70)
B and G (0.5 and 5 tenths)
C and F (5 hundredths and 0.05)
D and E (0.07 and 7 hundredths)

3 (a) 0.04, 0.11, 0.18, 0.20, 0.5
(b) 0.30, 0.34, 0.4, 0.59, 0.80
(c) 0.07, 0.21, 0.40, 0.54, 0.9
(d) 0.05, 0.1, 0.17, 0.3, 0.6

4 (a) 0.26, 0.31, 0.40, 0.5, 0.53
(b) 0.20, 0.26, 0.59, 0.6, 0.80
(c) 0.09, 0.2, 0.39, 0.9, 0.99
(d) 0.25, 0.3, 0.81, 0.87, 0.9

5 (a) 2.07, 2.39, 2.5, 2.8
(b) 1.01, 1.1, 1.43, 1.5

Section E (p 11)

1 (a) 3.27 m (b) 1.56 m
(d) 5.64 m (e) 4.98 m
(f) 2.09 m

2 A and H (1 metre and 8 cm, 1.08 m)
B and D (1.3 m, 130 cm)
C and G (1.8 m, 1 metre and 80 cm)
E and F (1 metre and 3 cm, 1.03 m)

3 (a) 2.09 m
(b) 3.20 m or 3.2 m
(c) 4.80 m or 4.8 m
(d) 9.01 m
(e) 6.40 m or 6.4 m
(f) 6.04 m

4

Name	Height in cm	Height in metres and cm	Height in metres
Mel	152 cm	**1 m 52 cm**	**1.52 m**
Ginger	**149 cm**	1 m 49 cm	1.49 m
Ali	**159 cm**	**1 m 59 cm**	1.59 m
Kay	**109 cm**	1 m 9 cm	**1.09 m**
Morag	**160 cm**	**1 m 60 cm**	1.6 m

5 (a) 2.87 m, 2.9 m, 3.08 m, 3.1 m
(b) 0.05 m, 0.5 m, 0.9 m, 1.01 m

6 B, C

Section F (p 12)

1 6.3, 6.17 and 6.37

2 (a) 4.09, 4.24, 4.25, 4.3, 5
(b) 0.08, 0.65, 0.7, 1, 1.5

3 TRIANGLE

4 6.06 and 6.48

5 (a) Jamie White
(b) Pete Smith
(c) Joe Hall (1.5 m)
Suneet Patel (1.48 m)
Garry Fulton (1.59 m)
Jamie White (1.6 m)

Section G (p 13)

1 (a) 700 or 7 hundreds
(b) 0.3 or 3 tenths
(c) 1
(d) 0.09 or 9 hundredths

2 A, S; B, R; C, T; D, U;
E, Q; F, P; G, R

3 (a) 63.7 (b) 72.7 (c) 62.9
(d) 38.74 (e) 33.74 (f) 39.64
(g) 39.73 (h) 30.1

4 (a) 458.1 + **10** = 468.1
(b) 34.91 + **0.01** = 34.92
(c) 103.6 + **50** = 153.6
(d) 75.32 – **1** = 74.32
(e) 54.8 – **0.8** = 54
(f) 587.5 + **400** = 987.5
(g) 6.77 – **0.7** = 6.07
(h) 6.77 – **0.07** = 6.7

8 Spot the ball

This unit introduces pupils to negative *x*-coordinates.
Negative *y*-coordinates are introduced in unit 14, 'Extending still further'.

Essential	Optional
Sheet 225 Squared paper Tracing paper, scissors, Newspaper sports pictures **Practice booklet** page 14	OHP transparency of sheet 225

Spot the ball (p 45)

T

Using negative coordinates only in the *x*-axis in the first instance should encourage the practice of always establishing the *x*-coordinate first.

◊ In question 1 pupils may at first give answers such as 'just below the racket'. Pupils should be led towards the idea of a grid being useful for them to describe their positions.

Now give pupils sheet 225 with the first quadrant grid imposed. Ask pupils how the grid could be extended to the left. This should easily lead to negative numbers being added to the *x*-axis. Get pupils to draw the rest of the cm square grid by careful measuring.

Pupils could then be asked to give the coordinates of their guesses, marking the guesses on a master grid on the board and/or on their own sheets. Copying the sheet on to an OHP transparency will help greatly.

Questions could then be asked such as who would have won if the ball had been at (4, 3), (5, 0), ($^-3$, 7), ($^-7$, 2) etc.

The pictures on the next page show where each ball actually is.

Design your own

Pupils could be asked to bring in their own pictures – most newspapers have at least two or three usable pictures of sporting activities every day. Each pupil should then trace the outline of the main player or players, and copy the outline on to squared paper.

Then get them to draw and label axes over their picture with (0, 0) in the middle of the bottom edge.

Each pupil then gets other pupils to guess the coordinates of where they think the ball is. The original photo can be used to decide who is nearest.

◊ Some pupils may have access to a scanner and simple photo-retouching packages and may wish to use these in designing their own game.

1

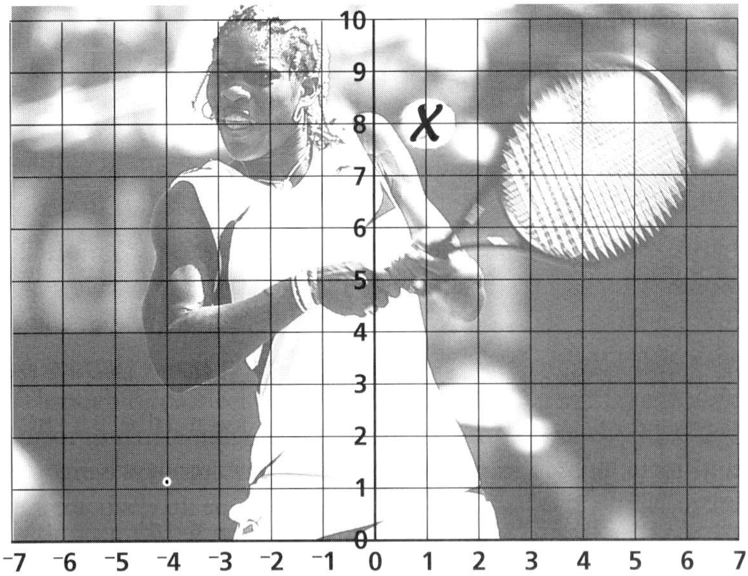

2

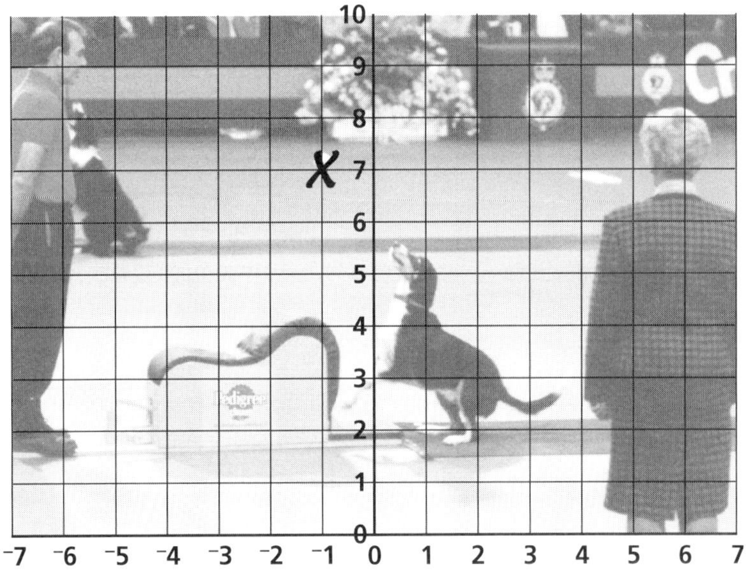

Practice booklet (p 14)

1

Ajit	(5, 8)	Beth	(5, 3)
Chloe	(3, 0)	Dave	(0, 6)
Elli	($^-4$, 3)	Fran	($^-2$, 2)
Glyn	($^-2$, 8)	Harry	($^-5$, 8)
Indu	($^-6$, 2)	Jack	($^-5$, 0)
Katie	($^-2$, 0)		

2 Beth's guess

3 (a) Dave (b) Elli (c) Elli

4

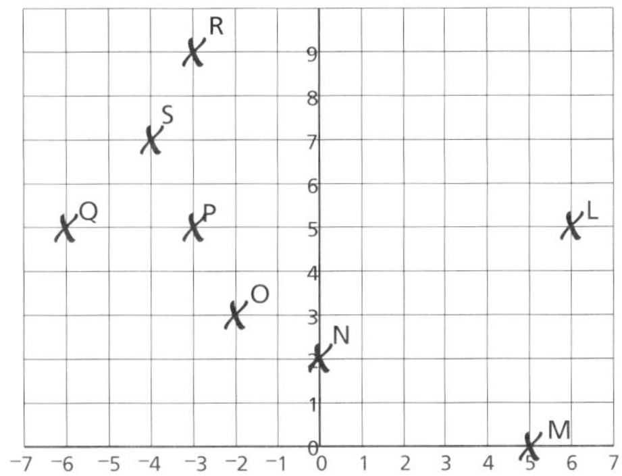

Review 1 (p 46)

1 (a) 100 (b) 130
(c) 60 (d) 240
(e) 180 mince pies, 210 crackers
100 streamers, 400 party poppers

2 (a) 7 (b) 11 (c) 21 (d) 201

3 (a) 7 (b) 12 (c) 8 (d) 27

4 (a) (2, 3) (b) (4, 4)
(c) (8, 6) (d) (2, 6)
(e) (i) (4, 7) (ii) (6, 8)
(iii) (8, 9)

5 4, 8, 12, 20, 28

6 1, 2, 3, 4, 5, 6, 10, 12, 15, 20

7 (a) 2 is a **factor** of 12
(b) 8 is a **multiple** of 4
(c) 24 is a **multiple** of 2
(d) 8 is a **factor** of 64

8 Three from:

2 rows of 60 3 rows of 40 4 rows of 30
5 rows of 24 6 rows of 20 8 rows of 15
(10 rows of 12 is given) 12 rows of 10
15 rows of 8 20 rows of 6 24 rows of 5
30 rows of 4 40 rows of 3 60 rows of 2.
(Sane caretakers would consider some of these arrangements impracticable.)

9 (a) 4.24 (b) 4.33
(c) 5.58 (d) 0.08

10 Gareth is taller by 59 centimetres.

11 1 m 1 cm, 1.09 m, 1.10 m, 128 cm, 1.5 m

12 (a) 234.57 (b) 234.66
(c) 244.56 (d) 334.56

13 (a) Abbi (1, 4), Ben (5, 3), Chaz (0, 2)
Dave ($^-2$, 5), Ewan ($^-5$, 1),
Fran ($^-1$, 0)
(b) Dave's guess was nearest.

Mixed questions 1 (Practice booklet p 15)

1 (a) 15 cups (b) 29

2 (a) (10, 3) (b) (12, 6)

3 (a) 24, 84, 120 (b) 2, 3, 6

4 15 by 1, 3 by 5, (and possibly 5 by 3 and 1 by 15) rectangles drawn

5 (a) 6.25 (b) 6.95 (c) 6.05

6 (a) 1.48 m (b) 2.05 m
(c) 4.3 m (d) 0.75 m

9 Hexathings

This unit revises and takes forward the work in 'Work to rule', unit 12 of *Book 1*.

In all sections, pupils should be encouraged to try to find the rule for a family of creatures in two ways:

- by looking at the geometry of the family
- by looking at the table of the number of red and white hexagons

p 48	**A** Creepy crawlies	Finding the rules for families of patterns in words
p 51	**B** Shorter rules	Writing rules in algebraic shorthand
p 52	**C** Chronic creatures	Linking shorthand rules to families of patterns

Essential

Sheet 226 (two or more per pupil)

Practice booklet pages 16 to 19

A Creepy crawlies (p 48)

Sheet 226

◊ In the initial discussion encourage pupils to look at the geometry of the patterns. They should see that for the caterpillars there is one white hexagon for each red one, plus another two.

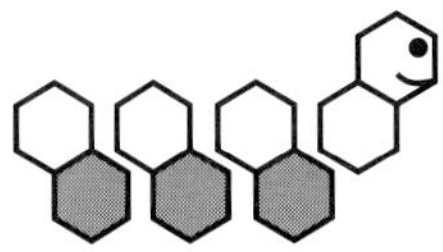

◊ Different patterns in this section require different orientations of sheet 226 and this needs pointing out to avoid confusion.

Invent your own family

Pupils could be given other rules to invent their own families for.

B Shorter rules (p 51)

'We returned to the written rules in A for examples.'

◊ When discussing the algebraic shorthand, emphasise that *w* (for example) does not stand for *white hexagons*, but for the ***number** of white hexagons*. Any letters can be used, of course.

C Chronic creatures (p 52)

'Throughout section C we stressed looking at the diagrams.'

C4 Pupils may decide that the rule is, say,
number of white hexagons = (*number of red hexagons* + 2) × 4,
leading to $w = 4(r + 2)$.

We have not emphasised multiplying a bracket by a number in the unit, but if this arises you may wish to discuss it with pupils.

A Creepy crawlies (p 48)

A1 (a) The pupil's caterpillar with 4 reds and 6 whites

(b) 6

A2 (a) The pupil's caterpillar with 6 reds and 8 whites

(b) 8

A3

reds	2	3	4	5	6
whites	**4**	**5**	**6**	**7**	**8**

A4 (a) 12 (b) 102

A5 number of white hexagons
= number of red hexagons + 2

A6 6

A7 (a) 10 (b) 12

A8 (a) The pupil's moth with 4 reds and 8 whites

(b) 8

A9 The pupil's moth with 2 reds and 4 whites

A10 20

A11 200

A12

reds	2	3	4	5	6	10	100
whites	**4**	**6**	**8**	**10**	**12**	**20**	**200**

A13 number of white hexagons
= number of red hexagons × 2

A14 (a) 16 (b) The pupil's moth

A15 (a) The pupil's earwigs with 1 red and 4 whites, 2 reds and 6 whites and 5 reds and 12 whites.

(b) 22 (c) 202

(d)

reds	1	2	3	4	5	6	10	100
whites	**4**	**6**	**8**	**10**	**12**	**14**	**22**	**202**

A16 number of white hexagons
= (number of red hexagons × 2) + 2

and number of white hexagons
= (number of red hexagons + 1) × 2

A17 (a) 18 (b) The pupil's earwig

Invent your own family

reds	2	3	4	5	6
whites	5	**6**	**7**	**8**	**9**

- The pupil's family to fit the table above

B Shorter rules (p 51)

B1 (a) $w = 3r + 2$ (b) $w = 2r + 2$

(c) $w = 2r + 3$ (d) $w = r + 3$

(e) $w = 3r$

B2 (a) $w = 4r + 2$ (b) $w = r + 2$

(c) $w = 5r$ (d) $w = 3r + 5$

B3 (a) 22 (b) 34

(c) 64 (d) 2

C Chronic creatures (p 52)

C1 (a) 13 (b) 31

(c) number of white hexagons = (number of red hexagons × 3) + 1

(d) $w = 3r + 1$ (e) 46

C2 (a) 14 (b) 104

(c) $w = r + 4$

C3 (a)

r	2	3	4	5	6	7
w	**7**	**9**	**11**	13	**15**	**17**

(b) 23 (c) 2003

(d) $w = 2r + 3$

C4 (a) 28 (b) 32

(c) 36 (d) 408

(e) $w = 4r + 8$

C5 Sea-snails $w = 3r + 2$
Sea-horses $w = r + 4$
Flatfish $w = 2r + 6$

What progress have you made? (p 55)

1 (a) The pupil's whelk with 4 reds and 10 whites

(b) The pupil's whelk with 6 reds and 14 whites

(c) 22

(d)

r	2	3	4	5	6	10	100
w	**6**	**8**	**10**	**12**	**14**	**22**	**202**

2 number of white hexagons = (number of red hexagons × 2) + 2

3 $w = 2r + 2$ or $w = 2(r + 1)$

Practice booklet

Section A (p 16)

1 (a) 10 whites (b) 7 whites

(c) The pupil's drawing with 4 spotty squares and 9 whites.

(d) The pupil's drawing with 6 spotty squares and 11 whites.

(e)

Number of spotty squares	2	3	4	5	6
Number of white squares	**7**	**8**	**9**	**10**	**11**

(f) 15 whites

(g) 105 whites

(h) number of white squares = number of spotty squares + 5

2 (a) 8 whites (b) 12 whites

(c) 6 whites

(d) The pupil's drawings with 4 grey squares and 6 grey squares

(e)

Number of grey squares	2	3	4	5	6
Number of white squares	**6**	**8**	**10**	**12**	**14**

(f) 22 white squares

(g) 202 white squares

(h) number of white squares = (number of grey squares × 2) + 2

Section B (p 18)

1 (a) $w = 3g + 1$ (b) $w = 2g + 3$

(c) $w = g + 5$ (d) $w = 3g$

(e) $w = 4g + 4$

2 (a) $w = 4g + 2$ (b) $w = g + 6$

(c) $w = 5g$

3 (a) 42 whites (b) 34 whites

(c) $w = 402$ (d) 3 greys

4 (a) 17 (b) 29 (c) 299

Section C (p 19)

1 (a) 19 whites (b) 37 whites

(c) 307 whites

(d) The pupil's explanation

(e) $w = 3g + 7$

2 (a) (i) 17 whites (ii) 21 whites

(iii) 25 whites

(b) 401 whites

(c) $w = 4g + 1$

10 Surfaces of solids

Pupils interpret nets of cubes and cuboids, and calculate surface areas.

T	p 56 **A** Cubes	Interpreting nets of cubes
T	p 58 **B** Cuboids	Nets and surface areas of cuboids

Essential	**Optional**
Sheets 227, 228 Scissors, sticky tape cm squared paper	Large version of net A on sheet 227 Sheet 229 Cardboard carton (for demonstration)
Practice booklet pages 20 and 21	

A Cubes (p 56)

T

'I started with a cube framework and asked what it would look like when "flattened out". Then I asked pupils to draw two different nets of a cube. A couple of pupils went wrong at first – but fine!'

The purpose of this section is to encourage visualisation. Encourage pupils to find solutions without cutting out nets. After they have found a solution, they may then cut out the nets to check their answers.

Sheets 227, 228, scissors, sticky tape

◊ You could start by demonstrating (or asking for a volunteer to demonstrate) how to fold and stick hexomino A to make a cube. A large copy of hexomino A might be useful for this.

Alternatively, you could start with a made-up cube and cut to make it into a net.

A1 If weaker pupils need to, they could cut out the hexominoes from sheet 227 and fold them to find out which ones make cubes. But they should cut them out only as a last resort.

A2, A3 Pupils who have already cut out the nets on sheet 227 for question 1 will need a second copy, so that they can colour the faces before cutting out.

***A7** You may need to point out that flaps that are not on adjacent edges on the net may end up on adjacent edges when the net is folded up.

B Cuboids (p 58)

Pupils interpret nets of cuboids and find surface areas.

Optional: Sheet 229, possibly enlarged on to A3 paper

Use the introduction to discuss nets of cuboids and also how to find the total surface area of a cuboid. You could do a demonstration, opening out a cardboard carton and discussing the areas of each of the faces.

B1 Sheet 229 is provided for those pupils who need to make the cuboids. To make the nets easier to handle it may be worth enlarging your copies on to A3 paper.

A Cubes (p 56)

A1 A, B, C and D are nets of a cube.

A2 The pupil's check

A3 The pupil's colouring of nets B, C and D

A4 R

A5

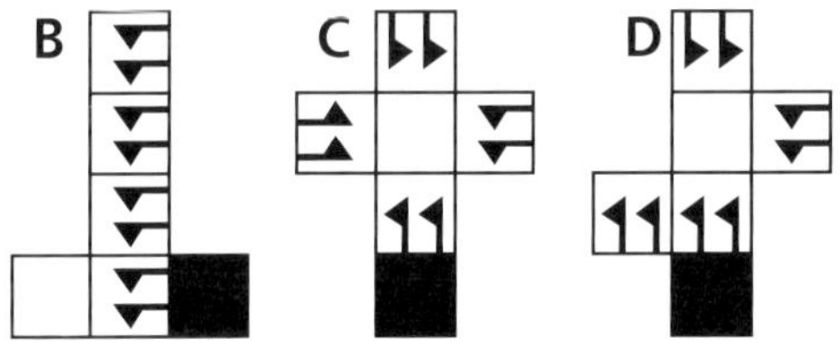

A6

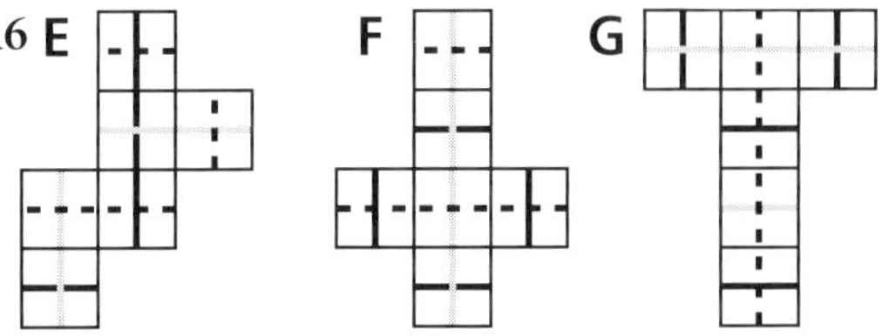

***A7** (a), (b), (e) and (f) are correct

B Cuboids (p 58)

B1 A is net of R; B is net of S;
C is net of P; D is net of Q

B2 The pupil's completed nets

B3 (a) (i) 12 cm^2 (ii) 12 cm^2
(b) 21 cm^2 (c) 28 cm^2
(d) 122 cm^2

B4 (a) 72 cm^2 (b) 78 cm^2 (c) 126 cm^2
(d) 128 cm^2 (e) 78 cm^2 (f) 234 cm^2
(g) 760 cm^2 (h) 9200 cm^2

What progress have you made? (p 60)

1 Q

2 B

3 1300 cm^2

Practice booklet

Section A (p 20)

1 A, D, E

Section B (p 21)

1 A belongs to R, B belongs to Q and C belongs to P.

2 The pupil's net

3 (a) 72 cm^2 (b) 68 cm^2
(c) 40 cm^2 (d) 280 cm^2

11 Lost for words

Pupils consolidate finding fractions of numbers in the context of word puzzles. Activity F2, 'What fraction of …', in unit 1 could usefully be done immediately before this activity.

Optional

Centimetre squared paper

◊ Pupils who find this activity difficult could write the words on centimetre squared paper with one letter in each square. Then shading the appropriate fraction of squares will identify the required letters.

For example, the first $\frac{1}{2}$ of SWEATY is SWE.

S	W	E	A	T	Y

Solve these puzzles (p 61)

1 SWEET

2 CLUE

3 ICE

4 PEACE

5 EXAM

6 SCHOOL

7 LESSON

8 CUBE

9 HOLIDAY

10 DIAMETER

Find the missing fraction (p 61)

1 $\frac{1}{2}$

2 $\frac{1}{4}$

3 $\frac{1}{3}$

4 $\frac{3}{4}$

12 Buying and selling

T	p 62 **A** Perfect pizzas	Multiplying to find the cost of items where weights are given as decimals
T	p 63 **B** Rounding to the nearest penny	Rounding amounts of money with three decimal places to the nearest penny
T	p 64 **C** Further rounding	Rounding to the nearest whole number and one decimal place

Practice booklet pages 22 to 24

A Perfect pizzas (p 62)

T

◊ You could start by using the illustration as a basis for some mental multiplication. For example:

- How much would it cost for 2 kg of olives?
- How much would 3 kg of cheese cost?

Now pose a question such as 'How much would it cost for 1.4 kg of cheese?' Discuss how you might estimate the answer to start with. For example, 1 kg of cheese costs £4.10 and 2 kg of cheese costs £8.20 so 1.4 kg must cost somewhere between £4.10 and £8.20. Ask pupils what calculation would give the exact answer. Then ask them to work out this exact answer using a calculator.

Ask similar questions about the other objects in the picture.

Include questions such as 'What is the cost of 3.4 kg of olives?' where the answer displayed on the calculator is 30.6 . Pupils need to realise that we write this as £30.60 and not as £30.6 (or £30.06).

◊ Many pupils find it difficult to grasp that multiplying by a number between 0 and 1 makes an amount smaller; you may need to spend some time on this. Pupils also need to appreciate that multiplying by 0.5 is the same as finding half or dividing by 2.

B Rounding to the nearest penny (p 63)

T

Pupils could work in groups and think about what they would pay for 2.4 kg of onions, 1.4 kg of carrots etc. They can then discuss what they think the costs should be and report their ideas to the rest of the class.

The first two examples (onions and carrots) give 1.248 and 0.462 on the calculator display. It may help pupils to think of, say, £1.248 as '£1.24 and a bit' which is between £1.24 and £1.25. Diagrams such as those below may help pupils to see that the amounts rounded to the nearest penny are £1.25 and £0.46.

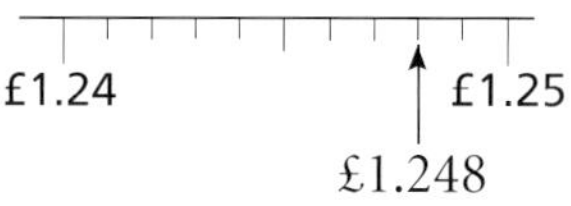

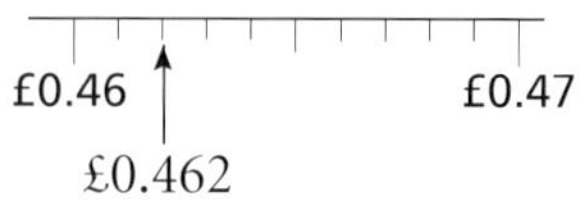

The next two examples (capers and onions) are of rounding up and down when the second digit after the decimal point is a zero. Some pupils find this confusing.

The pepperoni example gives £18.265, an amount exactly halfway between £18.26 and £18.27 where it is usual to round up to £18.27.

The final capers example gives £16.296 which many pupils find difficult. Often pupils write £16.210, having rounded the 9 up to a 10. Again, a diagram may help, thinking of £16.296 as between £16.29 and £16.30.

C Further rounding (p 64)

T

Pupils should have met rounding to the nearest whole number in *Book N*, but rounding to one decimal place may be new to them. You will probably need to discuss with them how to round (for example) 2.99 or 9.99 to 1 d.p. before they do the questions in this section.

A Perfect pizzas (p 62)

A1 (a) £10.25

(b)

Cheese	£10.25
Salami	£25.02
Olives	£14.95
Anchovies	£5.76
Tomatoes	£12.60
Mushrooms	£15.64

(c) £84.22

A2 (a)

Cheese	£7.38
Salami	£31.97
Olives	£6.50
Anchovies	£28.80
Mushrooms	£8.16
Tomatoes	£9.80
Total cost	£92.61

(b)

Cheese	£5.74
Salami	£41.70
Olives	£5.20
Anchovies	£9.60
Mushrooms	£0.85
Tomatoes	£0.70
Total cost	£63.79

A3 (a) (b) 2 kg of salami costs more (£27.80 as opposed to £17.00).

B Rounding to the nearest penny (p 63)

B1 (a) £6.71 (b) £2.68 (c) £4.24
(d) £10.34 (e) £4.73 (f) £1.41
(g) £7.14 (h) £1.20 (i) £2.42
(j) £3.30 (k) £1.20 (l) £3.00

B2 (a) £8.15 (b) £4.78 (c) £5.62
(d) £1.82 (e) £23.77 (f) £0.56
(g) £9.51 (h) £1.40 (i) £29.51
(j) £6.11 (k) £0.68 (l) £0.50

B3 (a)

Capers	£2.04
Pepperoni	£21.08
Carrots	£2.08
Onions	£2.81
Prawns	£11.24
Garlic	£0.90
Total cost	£40.15

(b)

Capers	£0.68
Onions	£1.09
Garlic	£2.09
Carrots	£1.12
Pepperoni	£35.13
Prawns	£26.23
Total cost	£66.34

C Further rounding (p 64)

C1 (a) 6.42 (b) 6.51 (c) 6.38
(d) 6.56 (e) 6.24

C2 (a) 6.4 (b) 6.5 (c) 6.4
(d) 6.6 (e) 6.2

C3 (a) 15 (b) 18 (c) 2 (d) 8
(e) 10 (f) 1 (g) 100 (h) 49
(i) 104 (j) 110

C4 (a) 3.7 (b) 4.1 (c) 110.7
(d) 0.2 (e) 9.6 (f) 0.2 (g) 9.9
(h) 6.1 (i) 10.1 (j) 12.9

C5 (a) 6.0 (b) 3.0 (c) 2.9 (d) 105.0
(e) 9.0 (f) 10.0

C6 (a) £31 (b) £63 (c) £94 (d) £188

***C7** (a) 38.6 grams (b) 57.8 grams
(c) 404.9 grams (d) 67.5 grams

What progress have you made? (p 65)

1 £38.35

2 £26.07

3 £8.40

4 (a) £3.92 (b) £6.74 (c) £5.91
(d) £4.20 (e) £1.97 (f) £8.60

5 (a) £12.89 (b) £3.13 (c) £1.23

6 (a) 45 (b) 6 (c) 11
(d) 0 (e) 26 (f) 40

7 (a) 1.22 (b) 1.25 (c) 1.29
(d) 1.16

8 (a) 1.2 (b) 1.3 (c) 1.3
(d) 1.2

9 (a) 5.9 (b) 6.1 (c) 9.0
(d) 0.3 (e) 0.1 (f) 20.0

Practice booklet

Section A (p 22)

1 (a) £4.18 (b) 60p or £0.60

2 £0.60 or 60p

3 £4.20

4 £1.96, £0.84, £1.52 Total cost = £4.32

5 (a)

Apples	£1.92
Figs	£2.32
Grapes	£2.28
Pears	£2.24
Watermelon	£4.20
Total	£12.96

(b)

Watermelon	£2.25
Apples	£3.12
Pears	£0.84
Figs	£8.70
Grapes	£3.04
Total	£17.95

6 £11.50 (Jan spends £14.50, Jo £3.00.)

Section B (p 23)

1 (a) £4.83 (b) £6.59 (c) £11.44
(d) £6.87 (e) £0.38 (f) £13.50
(g) £2.61 (h) £4.35 (i) £2.88
(j) £4.60 (k) £7.31 (l) £3.50

2 (a) £3.48 (b) £2.27 (c) £1.51
(d) £1.89 (e) £5.86 (f) £1.16
(g) £0.57 (h) £2.15 (i) £1.02
(j) £3.40

3 (a) £9.68 (b) £4.95 (c) £2.63
(d) £1.04 (e) £13.17 (f) £35.78

***4** £26.06

Section C (p 24)

1 (a) 3.62 (b) 3.69 (c) 2.04
(d) 2.12 (e) 1.79 (f) 1.85
(g) 6.85 (h) 6.98

2 (a) 3.6 (b) 3.7 (c) 2.0
(d) 2.1 (e) 1.8 (f) 1.9
(g) 6.9 (h) 7.0

3 (a) 6 (b) 9 (c) 6 (d) 14
(e) 11 (f) 13 (g) 8 (h) 10

4 (a) 7 (b) 2 (c) 10 (d) 11

5 (a) 7.3 (b) 2.5 (c) 3.6
(d) 4.9 (e) 9.6 (f) 10.1
(g) 20.0 (h) 100.0

6 (a) 8.7 (b) 2.6 (c) 5.0 (d) 10.0

7 (a) £77 (b) £129
(c) £52 (d) £258

***8** (a) 16.1 g (b) 32.3 g
(c) 40.4 g (d) 5.1 g

13 Percentages

This unit assumes no prior knowledge of percentages.

T	p 66 **A** How many per cent?	Identifying the percentage of a bar or a circle that is shaded
	p 67 **B** Easy percentages	Calculating 50% and 25% of money and numbers
T	p 68 **C** Shove the coin	Changing scores out of 10 or 5 to a percentage.

Essential	**Optional**
2p coins (one per pupil) A4 sheets of paper, compasses Sheets 230, 231 **Practice booklet** pages 25 to 27	Pie chart scales

A How many per cent? (p 66)

The aim is to introduce the per cent sign (%) and make pupils aware that percentages are commonly used in many different contexts. You should also get some idea of how much they already know about percentage.

Optional: Pie chart scales

T

◊ Ask pupils what, if anything, the percentages in the picture tell them (for example, that a 50% saving is half price). Pupils could be asked to bring in their own examples of labels or newspaper and catalogue pages which use percentages.

◊ It may be useful to point out that '%' is an abbreviation for '/100' or 'out of 100', although use of fractions is avoided in this unit.

◊ A pie chart scale may help in showing pupils what a circle divided into 100 looks like.

B Easy percentages (p 67)

C Shove the coin (p 68)

Finding percentages from results out of 5 and 10 is introduced in the context of a game with a coin.

2p coins or counters (one per pupil), a circle of radius 4 cm drawn on a sheet of paper (one per pupil), sheets 230, 231

T

'Pupils really enjoyed this game'

Shove the coin

The game is played with a piece of paper with a circle of radius 4 cm on it taped to a desk. Pupils then 'shove' the coins with the back of their hand. A success is getting the entire coin in the circle.

The circle needs to be taped down so that initially it is 10 cm away from the edge of the table. Then distances of 20, 30 and 40 cm can be tried. For each distance:

- Pupils have 10 tries and record their successes as ticks (✓) or crosses (✗).
- Pupils can calculate their percentage success using the table in C1 and record their results in the table in C2.
- They could also see what their percentage success rate was after 5 goes at each distance.

A How many per cent? (p 66)

A1 (a) 8%, 92% (b) 75%, 25%
(c) 50%, 50% (d) 49%, 51%
(e) 52%, 48%

A2 (a) 20%, 80% (b) 64%, 36%

B Easy percentages (p 67)

B1 (a) 5p (b) 50p (c) £30
(d) £250 (e) 37p (f) £1.20

B2 (a) £5 (b) £15 (c) £9
(d) £3.50 (e) £1.80 (f) £2.70

B3 (a) 5p (b) 25p (c) £20
(d) £125 (e) £25 (f) 30p

B4 (a) 100 g (b) 25 g (c) 150 g
(d) 60 g (e) 0.5 kg (f) 0.2 kg

B5 (a) 50% of £2 = £1
(b) 50% of £4.20 = £2.10
(c) 25% of £2.80 = 70p
(d) 25% of £12 = £3
(e) 50% of 50p = 25p
(f) 25% of £200 = £50

C Shove the coin (p 68)

C1 0%, 10%, **20%**, **30%**, **40%**, 50%, **60%**, **70%**, **80%**, **90%**, 100%

C2 The pupil's recorded results

C3 (a) 5 out of 10 = **50**%
(b) 7 out of 10 = **70**%
(c) 10 out of 10 = **100**%
(d) **3** out of 10 = 30%
(e) **9** out of 10 = 90%
(d) **2** out of 10 = 20%

C4 (a) 30% (b) 70%

C5 (a) 3 out of 5 = **60**%
(b) 1 out of 5 = **20**%
(c) 5 out of 5 = **100**%
(d) **4** out of 5 = 80%
(e) **2** out of 5 = 40%
(f) **0** out of 5 = 0%

C6 (a) 40% (b) 60% (d) 0%
(e) 80% (g) 40% (i) 100%

What progress have you made? (p 69)

1 69%

2 (a) 25p (b) £50 (c) 30 (d) 1.5 kg

3 (a) 70% (b) 30% (c) 60% (d) 20%

Practice booklet

Section A (p 25)

1 (a) 30%, 70% (b) 10%, 90%
(c) 40%, 60% (d) 23%, 77%
(e) 50%, 50% (f) 25%, 75%
(g) 50%, 50%

2 (a) 30% (b) 60% (c) 50%

Section B (p 26)

1 (a) A, C, F, H (b) E, G, I

2 15

3 21

4 (a) 20p (b) 4 cm (c) £1.80
(d) 0.70 m or 70 cm (e) £100
(f) £0.80 or 80p

5 (a) £3 (b) £6
(c) £2.40 (d) £18

Section C (p 27)

1 (a) 40% (b) 6 (c) 60%
(d) 70%

2 (a)

Score out of 5	0	1	2	3	4	5
Percentage	0%	20%	40%	60%	80%	100%

(b) 80%
(c) (i) 60% (ii) 20%

3 (a) 40% (b) 20% (c) 80%

4 (a) 60% (b) 60% (c) 50%

14 Extending still further

In 'Spot the ball' pupils met negative x-coordinates. Here they use points in all four quadrants for the first time.

Essential
Centimetre square paper
Practice booklet page 28

A Air traffic control (p 70)

T

Since pupils have already met negative x-coordinates, negative y-coordinates should need little introduction. Pupils could be shown a grid with negative x-coordinates and positive y-coordinates, and asked how this can be extended below.

◊ Pupils will need to draw a grid going from $^-3$ to $^+3$ on both axes to play the game. The game could be introduced at first by the class playing on the board, both suggesting and plotting points. Once the rules have been established pupils can then play it in pairs. The rules should ensure that partners check each other's plotting. A larger grid could be used but strategies become much less obvious.

◊ Pupils should be reminded that the x-coordinate is always given first.

◊ Pupils could investigate winning strategies for this game or find the maximum number of crosses that can be used on this grid (10). More able pupils could use this as a starting point for an open-ended investigation using different grids and rules.

B Going below (p 71)

B Going below (p 71)

B1 A (3, −1) B (1, −4) C (5, −2)
D (−3, −2) E (−4, −4) F (−4, 0)
G (0, −3)

B2 (a) 8 (b) 2 (c) 5
(d) 2 (e) 4 (f) 4

B3 The parallelogram and arrowhead have more than one solution. Only one is shown on this diagram.

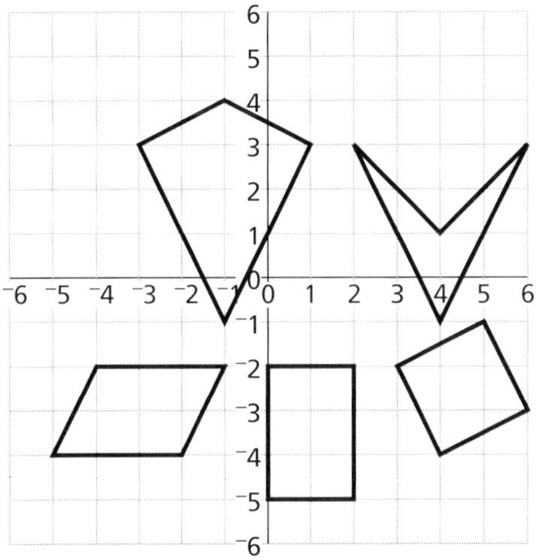

(a) (5, −1) (b) (0, −5)
(c) (−5, −4) or (−3, 0) or (1, −4)
(d) (−3, 3) (e) (2, 3) or (2, 1)

B4

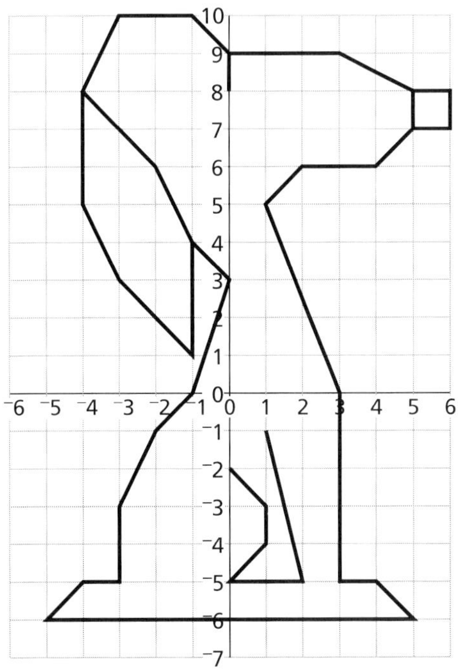

Practice booklet

Sections A and B (p 28)

1 A (1, 3) B (4, 0) C (3, −3)
D (−2, −2) E (−3, 0) F (−3, 2)
G (0, −3)

2 (a)

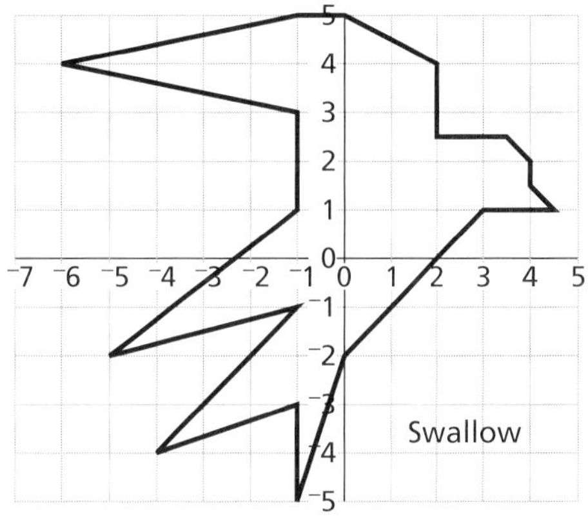

(b)

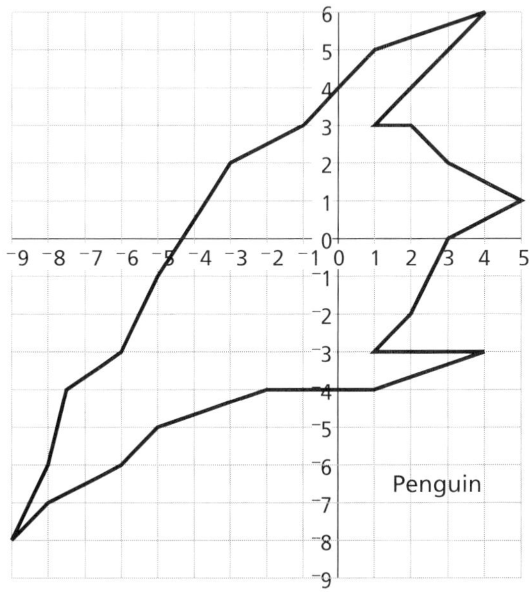

15 Numbers go to town

Pupils identify, continue and describe number patterns including square and triangle numbers.

Essential

Centimetre squared paper, coloured pens or pencils

Practice booklet pages 29 to 32

A Next door neighbours (p 72)

T

The pupils generate and explore patterns of consecutive sums.

◊ The activity could be started by asking pupils to think about the number sequencing in their street or block of flats.

◊ The sums on page 72 can be used to get the activity started.
For example, first ask what is 1 + 2 and 9 + 10 and write the whole sums on the board or OHP. Then invite pupils to come up with more two consecutive number sums. Write the sums pupils suggest at random places on the board. Then ask what patterns they can see. The aim is for pupils to arrive at the sequence of sums:
1 + 2 = 3
2 + 3 = 5
3 + 4 = 7
4 + 5 = 9 etc.

◊ Pupils could then look at the sums of 3, 4 or 5 consecutive numbers. If pupils suggest sums they can be written randomly as before but pupils should be led towards these being written in order so that the patterns can be seen.

1 + 2 + 3 = 6	1 + 2 + 3 + 4 = 10	1 + 2 + 3 + 4 + 5 = 15
2 + 3 + 4 = 9	2 + 3 + 4 + 5 = 14	2 + 3 + 4 + 5 + 6 = 20
3 + 4 + 5 = 12	3 + 4 + 5 + 6 = 18	3 + 4 + 5 + 6 + 7 = 25
4 + 5 + 6 = 15 etc.	4 + 5 + 6 + 7 = 22 etc.	4 + 5 + 6 + 7 + 8 = 30 etc.

◊ Ask pupils to look for and describe as many patterns as they can in these consecutive sums. They could work on this individually, in pairs or in small groups before coming together as a whole group to share their findings.

If asked, some pupils may be able to explain why the patterns arise. For example, they may see that in all the sequences of sums each sum can be obtained from the one above by adding 1 to each number on the left-hand side; this explains why the total goes up by the number of numbers on the left.

B Short houses, tall houses (p 72)

Pupils spot and continue number patterns based on multiples.

◊ Some of the patterns in this section are based on multiples but others are of the form nth term = $an + b$. Pupils will need help to see the rules such as 'multiples of 4 less 1'. It is advisable to let pupils complete question B1 and then discuss their results before moving on. Some pupils will simply count on, expressing their answers as 'goes up in fours' and may even need to draw out the continuing sequences of houses. Pupils should be encouraged towards more direct methods which avoid counting on (e.g. 'the tenth multiple of 4 less 1') but this approach will be dealt with more formally later.

C Where do the numbers live? (p 74)

In this section pupils look at the patterns formed by writing consecutive numbers in grids.

Centimetre squared paper

T

'This was a really good way of spotting that there are lots of different possible patterns.'

◊ Ask pupils to look for patterns within the grids on page 74. It is useful to see which multiples each grid is based on, that is:

2-row grid is based on multiples of 2 (or odd and even),
3-row grid is based on multiples of 3 etc.

Once pupils have had the opportunity to comment on these they could be asked questions such as

1 Which row is … in for each grid?
2 What number is on the left (or right) of … in each grid?
3 Give me three numbers between … and … in the …th row of the …-row grid.

D Squares (p 76)

T

'This was really enjoyed by pupils. One pupil branched out to different sized squares and tried to find a rule for each.'

This section looks at patterns involving square and triangle numbers.

Squared paper, coloured pens or pencils

◊ The first activity could be introduced using a blank copy of the 7 by 7 square displayed using your OHP. The aim is to leave as few single squares as possible.

B Short houses, tall houses (p 72)

B1 (a) (i) 10
(ii) The numbers on tall houses are multiples of 2.
(b) (i) 15
(ii) The numbers on tall houses are multiples of 3.
(c) (i) 25
(ii) The numbers on tall houses are multiples of 5.
(d) (i) 24
(ii) The numbers are 1 less than multiples of 5.

B2 (a) (i) 30 (ii) multiples of 3
(b) (i) 40 (ii) multiples of 4
(c) (i) 39
(ii) 1 less than multiples of 4
(d) (i) 48
(ii) 2 less than multiples of 5

B3 (a) (i) 32 (ii) multiples of 4
(b) (i) 64 (ii) multiples of 8
(c) (i) 48 (ii) multiples of 6
(d) (i) 48 (ii) multiples of 6
(e) (i) 160 (ii) multiples of 20
(f) (i) 47 (ii) 1 less than multiples of 6

B4 (a) No
(b) The numbers on tall houses are multiples of 3, which 31 is not.

***B5** (a) No
(b) The numbers on tall houses are 1 less than multiples of 4, and 48 is a multiple of 4.

B6 (a) 15, 19 (b) 12, 14 (c) 14, 17
(d) 22, 27 (e) 16, 19 (f) 6, 18
(g) 4, 8 (h) 2, 6 (i) 5, 9
(j) 9, 21 (k) 16, 22 (l) 10, 11
(m) 5, 20 (n) 2, 3

C Where do the numbers live? (p 74)

C1 2-row grid
(a) Row 1
(b) Odd numbers are in row 1 and 37 is odd.

3-row grid
(a) Row 1
(b) The grid is based on multiples of 3 and row 1 has numbers which are 1 more than multiples of 3.

4-row grid

(a) Row 1

(b) The grid is based on multiples of 4 and numbers which are 1 more than a multiple of 4 are in row 1.

5-row grid

(a) Row 2

(b) The grid is based on multiples of 5 and numbers which are 2 more than a multiple of 5 are in row 2 **or** all numbers ending in 2 or 7 are in this row.

C2 (a) **54** 59 **64**
5 less than and 5 greater than 59

(b) **77** 82 **87**
5 less than and 5 greater than 82

(c) **96** 101 **106**
5 less than and 5 greater than 101

(d) The units digits are the same.

C3 (a) The two columns are

31	36
32	37
33	38
34	39
35	40

(b) Odd numbers are yellow, even are red.

C4 (a) Row 2 (b) Row 3
(c) Row 5 (d) Row 4

C5 Any three of 74, 79, 84, 89

***C6** (a) In red rows the units digits (odd) follow this cycle

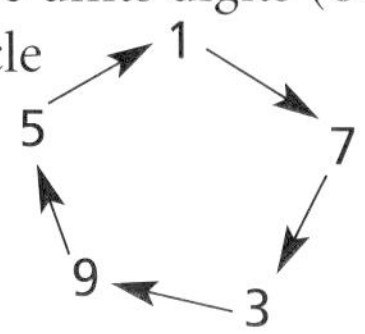

In yellow rows the units digits (even) follow this cycle

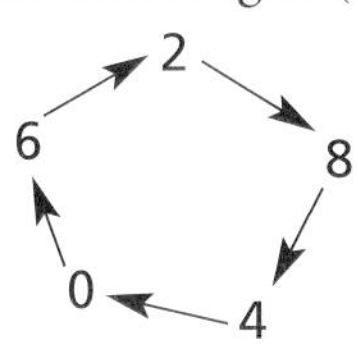

(b) Odd numbers are in red rows, even numbers in yellow.

(c) (i) Row 6 (as a multiple of 6)
(ii) Row 3 (as 3 more than a multiple of 6)

(d) 51 57 63
or 57 63 69
or 63 69 75

(e) 50 56 62
or 56 62 68
or 62 68 74
or 68 74 80

D Squares (p 76)

D1 (a) 1 (b) 4 (c) 9 (d) 16
(e) 25 (f) 36 (g) 49 (h) 64
(i) 81 (j) 100

D2 64, 36, 100, 144, 121

D3 (b) 1 3 6 10 15

(c) The pupil's triangle patterns with 21 and 28 dots

(d) 36 and 45 (e) 66

D4 Square numbers

D5 (a), (b)

1 = **1**
1 + 2 + 1 = **4**
1 + 2 + 3 + 2 + 1 = **9**
1 + 2 + 3 + 4 + 3 + 2 + 1 = **16**
1 + **2** + **3** + **4** + **5** + **4** + **3** + **2** + **1** = **25**
1 + **2** + **3** + **4** + **5** + **6** + **5** + **4** + **3** + **2** + **1** = **36**

(c) Square numbers

(d) 64

D6 (a)

1 = **1**
1 + 3 = **4**
1 + 3 + 5 = **9**
1 + 3 + 5 + 7 = **16**
1 + **3** + **5** + **7** + **9** = **25**
1 + **3** + **5** + **7** + **9** + **11** = **36**

The answers are square numbers.

(c) $(1 \times 2) \div 2 = \mathbf{1}$
$(2 \times 3) \div 2 = \mathbf{3}$
$(3 \times 4) \div 2 = \mathbf{6}$
$(\mathbf{4} \times \mathbf{5}) \div 2 = \mathbf{10}$
$(\mathbf{5} \times \mathbf{6}) \div 2 = \mathbf{15}$
The answers are triangle numbers.

(d) $(1 \times 2) - 1 = \mathbf{1}$
$(2 \times 3) - 2 = \mathbf{4}$
$(3 \times 4) - 3 = \mathbf{9}$
$(\mathbf{4} \times \mathbf{5}) - \mathbf{4} = \mathbf{16}$
$(\mathbf{5} \times \mathbf{6}) - \mathbf{5} = \mathbf{25}$
The answers are square numbers.

What progress have you made? (p 78)

1 (a) 15, 18 (b) 13, 16 (c) 36, 42
(d) 21, 25 (e) 138, 145 (f) 37, 50

2 The even rows are multiples of 2, the odd rows are the odd numbers.
The second column is multiples of 3 with other columns being 2 more or 2 less.
The diagonals are multiples of 5 with constant numbers added on.

3 (a) $(2 \times 2 - 2) \div 2 = \mathbf{1}$
$(3 \times 3 - 3) \div 2 = \mathbf{3}$
$(4 \times 4 - 4) \div 2 = \mathbf{6}$
$(\mathbf{5} \times \mathbf{5} - \mathbf{5}) \div \mathbf{2} = \mathbf{10}$
$(\mathbf{6} \times \mathbf{6} - \mathbf{6}) \div \mathbf{2} = \mathbf{15}$
The answers are triangle numbers.

(b) $(1 \times 3) - (1 \times 2) = \mathbf{1}$
$(2 \times 5) - (2 \times 3) = \mathbf{4}$
$(3 \times 7) - (3 \times 4) = \mathbf{9}$
$(4 \times 9) - (4 \times 5) = \mathbf{16}$
$(\mathbf{5} \times \mathbf{11}) - (\mathbf{5} \times \mathbf{6}) = \mathbf{25}$
$(\mathbf{6} \times \mathbf{13}) - (\mathbf{6} \times \mathbf{7}) = \mathbf{36}$
The answers are square numbers.

Practice booklet

Section B (p 29)

1 (a) (i) 2, 4, 6 (ii) 12
(b) (i) 3, 7, 11 (ii) 23
(c) (i) 1, 4, 7 (ii) 16
(d) (i) 2, 7, 12 (ii) 27

2 (a) 3, 6, 9 (b) Yes
(c) 45 is a multiple of 3.

3 (a) Row A 5, 11, 17
Row B 4, 12, 20
(b) 23 (c) 28

4 (a) 3, 6, 9, **12**, **15**
(b) 4, 7, 10, **13**, **16**
(c) **8**, **10**, 12, 14, 16
(d) 5, **7**, 9, **11**, 13
(e) 6, 10, **14**, **18**, 22
(f) 3, 8, 13, **18**, 23, **28**

Section C (p 30)

1 (a) 4, 7, 10, 13, 16, **19**, **22**, **25**, **28**, **31**
(b) The numbers go up in 3s.
(c) 40 and 46
(d) 9, 13, 17, **21**, **25**, **29**, **33**, **37**, **41**, **45**
(e) The numbers go up in 4s.
(f) No
(g) All the numbers are odd and 60 is even.

2 (a)

5	8	11	**14**
6	9	**12**	15
7	**10**	**13**	16
8	11	14	**17**

(b) (i) Row 2 (ii) Row 4 (iii) Row 1
(c) 50 and 56

Section D (p 31)

1 (a) $\mathbf{1} + \mathbf{9} = 10$ (b) $\mathbf{4} + \mathbf{16} = 20$
(c) $\mathbf{1} + \mathbf{4} + \mathbf{25} = 30$ (d) $\mathbf{25} + 11 = \mathbf{36}$
(e) $\mathbf{36} - \mathbf{25} = 11$ (f) $13 + \mathbf{36} = \mathbf{49}$
(g) $\mathbf{9} + \mathbf{16} = \mathbf{25}$
(h) $1 \times 1 = 1$ $1 \times 4 = 4$
$1 \times 9 = 9$ $1 \times 16 = 16$
$1 \times 25 = 25$ $1 \times 36 = 36$
$1 \times 49 = 49$ $1 \times 64 = 64$
$4 \times 4 = 16$ $4 \times 9 = 36$
$4 \times 16 = 64$

2 (a) $4 \times 1 + 5 = \mathbf{9}$

$5 \times 2 + 6 = \mathbf{16}$

$6 \times 3 + 7 = \mathbf{25}$

$\mathbf{7} \times \mathbf{4} + \mathbf{8} = \mathbf{36}$

$\mathbf{8} \times \mathbf{5} + \mathbf{9} = \mathbf{49}$

The answers are square numbers.

(b) $2 + 1 = \mathbf{3}$

$3 + 2 + 1 = \mathbf{6}$

$4 + 3 + 2 + 1 = \mathbf{10}$

$\mathbf{5} + \mathbf{4} + \mathbf{3} + \mathbf{2} + \mathbf{1} = \mathbf{15}$

$\mathbf{6} + \mathbf{5} + \mathbf{4} + \mathbf{3} + \mathbf{2} + \mathbf{1} = \mathbf{21}$

The answers are triangle numbers.

(c) $(1 \times 3) + 1 = \mathbf{4}$

$(2 \times 4) + 1 = \mathbf{9}$

$(3 \times 5) + 1 = \mathbf{16}$

$(\mathbf{4} \times \mathbf{6}) + \mathbf{1} = \mathbf{25}$

$(\mathbf{5} \times \mathbf{7}) + \mathbf{1} = \mathbf{36}$

The answers are square numbers.

(d) $2 \times 2 - 1 \times 1 = \mathbf{3}$

$3 \times 3 - 2 \times 2 = \mathbf{5}$

$4 \times 4 - 3 \times 3 = \mathbf{7}$

$5 \times 5 - 4 \times 4 = \mathbf{9}$

$6 \times 6 - 5 \times 5 = \mathbf{11}$

The answers are the odd numbers.

3 (a), (b)

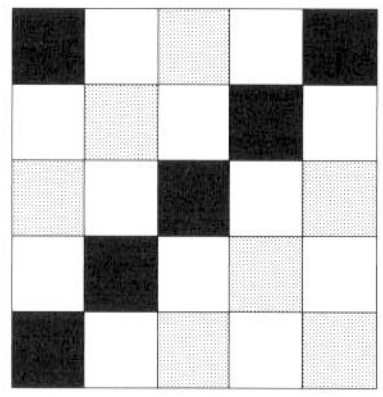

$1 + 2 + 3 + 4 + 5 + 4 + 3 + 2 + 1 = 25$

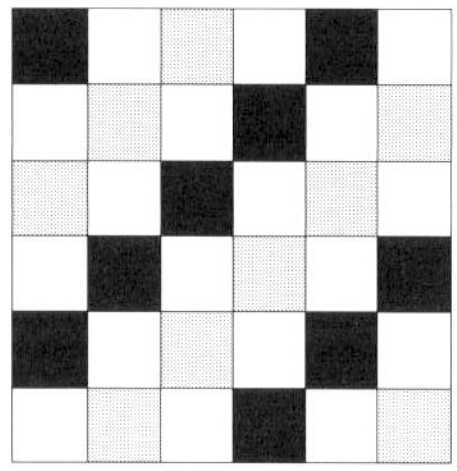

$1 + 2 + 3 + 4 + 5 + 6 + 5 + 4 + 3 + 2 + 1 = 36$

(c) $1 + 2 + 3 + 4 + 5 + 6 + 7 + 6 + 5 + 4 + 3 + 2 + 1 = 49$

(d) $1 + 2 + 3 + 4 + 5 + 6 + 7 + 8 + 9 + 10 + 9 + 8 + 7 + 6 + 5 + 4 + 3 + 2 + 1 = 100$

Review 2 (p 79)

1 (a) 4 whites

(b)

r	4	5	6	7	8
w	**3**	**4**	**5**	**6**	**7**

(c) $w = r - 1$

2 (a) No (b) No (c) Yes

3 (a) $94\,\text{cm}^2$ (b) $280\,\text{cm}^2$ (c) $132\,\text{cm}^2$

4 (a) KITTEN (b) PUPPY
(c) CAT (d) HORSE
(e) DUCKLING

5 (a) £13.97 (b) £0.58 (c) £8.03

6 (a) 60 g (b) 20 kg (c) £2.40

7 (a) 60% (b) 40%

8 (a) A $(^-4, ^-2)$, B $(^-2, ^-1)$, C $(^-1, 1)$,
D $(2, 1)$, E $(2, ^-2)$

(b)

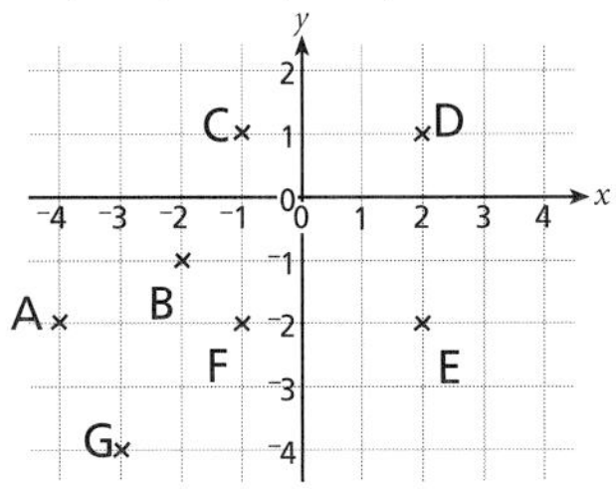

(c) A square (d) A kite

9 (a) 60 (b) The pupil's explanation

10 With the pupil's explanations
(a) 90 (b) 29 (c) 58 (d) 76

11 (a) 1, 9, 15 and 25
(b) 1, 4, 9, 16, 25 and 64
(c) 4, 12, 16, 32 and 64
(d) 1, 9, 15

Mixed questions 2 (Practice booklet p 33)

1 (a) 14 whites

(b)

g	1	2	3	4	5	6	10	100
w	**6**	**8**	**10**	**12**	**14**	**16**	**24**	**204**

(c) $w = 2g + 4$

2 (a)

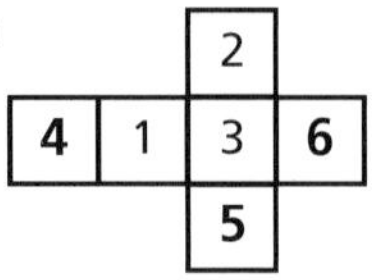

(b)

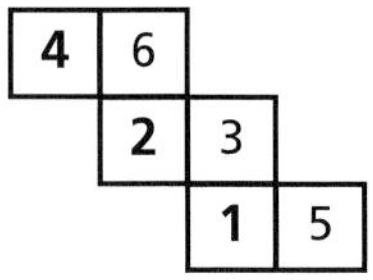

(c)

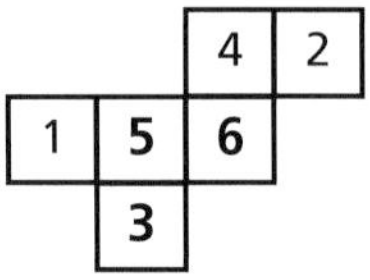

3 (a) £1.82 (b) £7.67 (c) £3.37

4 (a) (i) 25% (ii) 75%
(b) (i) 40% (ii) 60%
(c) (i) 70% (ii) 30%

5 (a) 14 m (b) £10
(c) 2.5 kg (d) 4 cm

6 (a) A $(^-3, 1)$, B $(^-2, 0)$, C $(1, 3)$,
D $(3, 2)$, E $(2, 1)$, F $(2, ^-2)$,
G $(0, ^-1)$, H $(^-2, ^-4)$, I $(^-3, ^-2)$

(b) (i) Rectangle (ii) Parallelogram
(iii) Kite

7 (a) $1 \times 3 - 1 \times 2 = \mathbf{1}$
$2 \times 4 - 2 \times 2 = \mathbf{4}$
$3 \times 5 - 3 \times 2 = \mathbf{9}$
$\mathbf{4} \times \mathbf{6} - \mathbf{4} \times \mathbf{2} = \mathbf{16}$
$\mathbf{5} \times \mathbf{7} - \mathbf{5} \times \mathbf{2} = \mathbf{25}$
The answers are square numbers.

(b) $(1 \times 1 + 1) \div 2 = \mathbf{1}$
$(2 \times 2 + 2) \div 2 = \mathbf{3}$
$(3 \times 3 + 3) \div 2 = \mathbf{6}$
$(\mathbf{4} \times \mathbf{4} + \mathbf{4}) \div \mathbf{2} = \mathbf{10}$
$(\mathbf{5} \times \mathbf{5} + \mathbf{5}) \div \mathbf{2} = \mathbf{15}$
The answers are triangular numbers.

8 (a) $268\,\text{cm}^2$ (b) $224\,\text{cm}^2$

16 Mean

The concept is introduced through the idea of 'equal sharing' of concrete objects. Practical work is included. Pupils work with decimals but in general the questions avoid the issue of rounding or truncating.

A spreadsheet or graphic calculator can be used to calculate the mean of a list of data. Pupils could complete part of this unit using a spreadsheet or graphic calculator.

p 81	**A** Fair's fair	The mean of whole-number data, where the mean is a whole number or has one decimal place
p 85	**B** Comparing means	A practical activity involving larger sets of data
p 85	**C** Decimal means	Means of data involving decimals

Essential	**Optional**
5 or 10 packets of Smarties, M&Ms etc. OHP transparencies of sheets 232 and 233 Sheet 234 on thick card (one per pair)	Tiles, counters or multilink

Practice booklet pages 35 to 38

A Fair's fair (p 81)

The objective of the teacher-led discussion is to introduce the mean for a set of discrete data through the idea of 'equal sharing'.

5 or 10 packets each of Smarties, M&Ms etc for 'Pack it in'
Optional: Tiles, counters or multilink to represent the cans/bottles

◊ Cans are used so pupils can visualise them being shared out equally between members of the groups. Introduce the mean number of cans as the result of sharing the cans equally between members of the group. Using tiles, counters or multilink to represent the cans may help.

◊ The point of the discussion is to arrive at a mean of $20 \div 4 = 5$ cans for Ann's group and $18 \div 3 = 6$ cans for Ken's group. Ken's group has the highest mean number of cans per person and so can be judged to have done 'better'. This may be judged to be unfair by some pupils who still judge that Ann's group did better because they collected more cans.

◊ Pupils could try questions A1 to A4 without a calculator to consolidate number skills. Use of a calculator is assumed in all subsequent work.

A5 Decimals are avoided in questions A1 to A4. You might want to mention them in your introduction or wait until question A5. Note that group 5 has a mean which is a recurring decimal. You may need to discuss how to deal with this.

Pack it in (p 84)

T

'The pupils enjoyed this and we had a really good discussion.'

You will need 5 or 10 small packets of M&Ms, 5 or 10 tubes of Smarties and 5 or 10 packets of other sweets. It is less expensive (but more boring) to count the contents of reuseable packets such as paper clips, drawing pins etc.

◊ Discuss what 'average contents' means – the mean number of sweets in the packets.

Divide the class into groups, and give each group one of the packets to count the contents. When all groups have finished, results can be recorded on the board, and the mean contents of the packets worked out. (Counting the contents of 5 or 10 packets avoids the complication of non-terminating decimals.)

Pupils can now discuss what a sensible 'average contents' label for the packets of M&Ms might be.

◊ The activity can be repeated with the different types of sweets.

B Comparing means (p 85)

T

'Items were attached to the board and we compared the average short-term memory of boys against girls. They really enjoyed this – so much so that they wanted to test how well they'd remembered the items in the next lesson so we did their average long-term memory too!'

Pupils find the mean for longer lists of discrete data where they are not explicitly asked to find the totals before calculating the mean.
They use means to compare sets of data.

The context is memory experiments. Pupils are usually highly motivated by calculating means about themselves.

OHP transparencies of sheets 232 and 233

◊ Explain to pupils that they are going to carry out an experiment to compare the ability to remember pictures and words. Start with a transparency of sheet 232. Show it to pupils for 12 seconds and then give them approximately 45 seconds to write down as many objects as they can remember. (Order and spelling do not matter.) Now do the same with sheet 233.

For each experiment, find the mean number of items remembered. These means may be non-terminating decimals which will need rounding to one decimal place. Pupils may need reminding about rounding.

B1, B2 Mistakes are less likely if pupils work in pairs with one reading the list of numbers to the other. All figures in questions B1 and B2 are based on results of actual experiments.

C Decimal means (p 85)

Pupils find means for sets of continuous data where both the data and the mean are decimals (one or two decimal places). The context for this section is weights (of babies and seals).

◊ The information on baby weights is based on authentic data.

The UK means for 1994/95 were 3.378 kg (boys) and 3.264 kg (girls) based on all deliveries (single births, twins, triplets etc).

◊ All names and weights for the seals are based on authentic data obtained from The National Seal Sanctuary in Gweek, Cornwall. Information on The National Seal Sanctuary including recent rescues can be found at www.sealsanctuary.co.uk

C4 You may like to point out that although the mean weight for the boys is higher, it is a girl (Phuong) who is the heaviest baby in these tables.

***C12** These figures give non-terminating decimals as mean weights.

Aiming high (p 87)

This game consolidates finding the mean for discrete data where the mean can be a whole number or a decimal.

Sheet 234 (copied on to thick card – one for each pair)

◊ The cards can be placed in a pile and the top card picked each time or they can be spread out for pupils to choose from. Either way, they need to be copied on to card that is thick enough so that the numbers do not show through.

Alternatively, the game can be played in groups of three, with one player acting as the dealer (as in 'Pontoon' with players 'twisting' or 'sticking').

◊ In one school, pupils organised a class tournament where 2 points were awarded for a win, 1 for a draw and 0 for a loss.

◊ Pupils can be encouraged to think about a strategy for maximising their chances of winning each time.

'This game really helped emphasise that you need to divide by the number of pieces of data.'

'I thought pupils would have problems deciding when one decimal was greater than another but they coped pretty well.'

A Fair's fair (p 82)

A1 (a) 28 cans (b) 7 cans

A2 Amit's group: 3 cans
Jason's group: 4 cans

A3 (a) Sarah's group: 6 cans
Wayne's group: 4 cans
Sohan's group: 3 cans
Ravi's group: 8 cans

(b) Ravi's group

(c) Sohan's group

A4 (a) Group 5 (b) Group 2
(c) Group 4 (d) Group 1
(e) Group 2 (f) Group 3

A5 (a) Group 2: 11.4 bottles
Group 3: 8.5 bottles
Group 4: 12.2 bottles
Group 5: 6.6 or 6.7 bottles
(b) Group 4 (c) Group 5

B Comparing means (p 85)

B1 (a) $142 \div 20 = 7.1$ pictures
(b) $195 \div 25 = 7.8$ pictures
(c) Class 10T has the higher mean.

B2 (a) 7Y: $126 \div 20 = 6.3$ words
10T: $180 \div 25 = 7.2$ words
(b) Class 10T has the higher mean.

***B3** For each class, the mean number of pictures is higher than the mean number of words. This suggests both classes are better at remembering pictures.

C Decimal means (p 85)

C1 (a) Ellen
(b) Rosanna, Ruth, Poppy, Jenny
(c) $19.56 \div 6 = 3.26$ kg

C2 (a) $18.1 \div 5 = 3.62$ kg
(b) Daniel, Andrew, Peter

C3 (a) $23.74 \div 8 = 2.97$ kg
(b) $24.32 \div 8 = 3.04$ kg

C4 The pupil's comments, for example:

The mean weight for the boys is higher than the mean weight for the girls in C3.

The mean weights for the baby girls in C3 (born two weeks early) is less than the mean weight in C1 (c).

The mean weights for the baby boys in C3 (born two weeks early) is less than the mean weight in C2 (a).

C5 65 kg

C6 Peace, Shanny, Montague, Allis

C7 Montague

C8 55.2 kg

C9 Yorkie, Gurnard, Montague, Bass

C10 $172.3 \div 10 = 17.23$ kg

C11 $690 \div 10 = 69$ kg

***C12** (a) $1315 \div 6 \approx 219$ kg or 219(.166...) kg
(b) $1125 \div 7 \approx 161$ kg or 160(.714...) kg

What progress have you made? (p 88)

1 (a) $28 \div 4 = 7$ cans
(b) $30 \div 5 = 6$ cans

2 $452 \div 5 = 90.4$ matches

3 (a) $83.2 \div 8 = 10.4$ kg
(b) $70 \div 7 = 10$ kg

Practice booklet

Section A (p 35)

1 (a) 25 children (b) 5

2 (a) (i) £5 (ii) £4 (iii) £3
(b) The Lings had the highest mean.
(c) The Cassells had the lowest mean.

3 A Scruton B Ling
C Instone D Scruton
E Cassell F Shilton

Section B (p 37)

1 (a)

Event	Total	Mean
Fun run	630	105
Sponsored silence	522	104.4
Sponsored spell	422	105.5
Sponsored walk	503	100.6

(b) The mean was lowest for the sponsored walk.

2 The pupil's answer explained, for example:
- Sponsored spell as it has the highest mean
- Sponsored silence as the numbers taking part are going up.

Section C (p 38)

1

Street	Total (£)	Mean (£)
Gresham St	18.72	3.12
Nonesuch Rd	23.45	3.35
Sidcup St	13.60	3.40
Bassett Cres	15.30	3.06

(a) Nonesuch Road

(b) Sidcup Street

(c) The pupil's answer explained.

***2** (a) Mean = £10 ÷ 3 = £3.3333…
= £3.33 to the nearest penny

(b) The mean for High Street was higher than the means for Gresham Street and Bassett Cresent, but lower than those for Sidcup Street and Nonesuch Road.

17 To scale

This unit first deals with interpreting drawings where the scale is drawn on the map or diagram, and then introduces scales expressed in the form '1 cm stands for ... '.

T	p 89 **A** Cutting down to size	Comparing real-life objects with diagrams
	p 90 **B** Using scales	Using scales given on a diagram
T	p 93 **C** One to one	Using scales where 1 cm stands for 1 m, or 1 km
	p 95 **D** Different scales	Using simple scales of the type '1 cm stands for ...'

Essential	**Optional**
Tape measures	Transparency of sheet 235
Scissors	Long tape measures
String or cotton	
Sheets 235–240	
Practice booklet pages 39 to 41	

A Cutting down to size (p 89)

This activity is intended to stimulate discussion on the need for scale drawings and why different scales are needed.

Tape measures, sheet 235
Optional: transparency of sheet 235

T

◊ A transparency of sheet 235 can be projected on to a wall so that Robert Wadlow's image is 208 cm tall. Pupils can then cut Robert out from sheet 235 and see how many of these fit into the real height (there should be 26). They can also stand against the projected image and compare their own height directly with his. Alternatively a life-size outline of Robert Wadlow could be pinned up.

Pupils can then measure their own heights and draw themselves on copies of sheet 235.

◊ In 'How many times larger' it is not necessary to find precise scale factors, simply to see that there is a range of scales used in these similar size sketches.

Rough answers might be: ruler, 5 times; shoe, 10 times; cat, 8 times; car, 100 times.

B Using scales (p 90)

Sheets 236–238 Optional: string or cotton

This section provides a range of examples with different scales that can be used to read from maps and diagrams. Scrap paper can be used to measure (in some cases string may be easier) and transferred to the scale.

B2(c) Using a piece of string to follow a likely route would be more realistic, and is a useful practical skill.

B5 String might be useful to work out the distance round the island.

C One to one (p 93)

This section uses simple scales where a centimetre stands for a metre or kilometre.

Sheets 239 and 240, scissors Optional: transparency of sheet 239, string

How far between the Snorings?

You can use sheet 239 to ask additional questions of the types given on the pupil's page. The answers to these are:

Distance between Great and Little Snoring	2 km (to nearest km)
Length of railway line from Wells to Walsingham	6 km
The village 6 km due west of Wighton	New Holkham

C4 Pupils may wish to use a piece of string or edge of a sheet of paper, as in earlier work, to measure distances.

D Different scales (p 95)

◊ The key points to raise in the initial activity are:

- To convert a measurement from a plan into a 'real' distance you need to multiply by the scale factor.
- To convert a measurement from a 'real' distance to one on the plan you need to divide by the scale factor.

Later questions may need a calculator. If the arithmetic proves too difficult, pupils could make their own 'scales' of the type used previously rather than convert measurements using scale factors.

T

B Using scales (p 90)

B1 (a) 21 m (b) 6 m
(c) 60 m (d) 64 m

B2 (a) (i) 150 km (ii) 300 km
(iii) 550 km
(b) Yes – Liverpool to Dublin is less than 250 km.
(c) 1000 km

B3 (a) 50 km (b) 40 km
(c) It is further to Zeebrugge by 20 km.

B4 (a) Free (b) £5
(c) £5 (d) £10

B5 About 51 to 52 miles

B6 The pupil's cycle route

B7 (a) 0.6 m (b) 1.2 m
(c) 1.4 m (d) The pupil's design

C One to one (p 93)

C1 (a) 9 km (b) 5 km
(c) 12 km (d) 3 km

C2 5.9 km to 6 km

C3 (a) Bircham Tofts (b) North Creake
(c) Syderstone (d) Burnham Market

C4 Pupils answers may vary, but roughly
Leg A: 11 km
Leg B: 19 km
Leg C: 7 km
Leg D: 11 km
Leg E: 12 km
Total: 60 km

C5 (a) Roughly 8 m
(b) Roughly 3 m by 2 m (c) 7 m
(d) (i) Yes (ii) No (iii) No
(e) 2 m by 5 m (f) 10 m^2

C6 (a) Only the car can get round.
(b) The pupil's roundabout design

D Different scales (p 95)

D1 (a) 32 m (b) 20 m (c) 30 m

D2 (a) Yes (b) No

D3 (a) 330 m (b) 44 s

D4 (a) (i) 8 m (ii) 20 m
(iii) 40 m (iv) 10 m
(b) 3 cm, 12 m (c) 8 cm, 32 m
(d) $15\frac{1}{2}$ cm, 62 m

D5 (a) 9 cm (b) 450 m (c) 2 cm
(d) 3 cm (e) $6\frac{1}{2}$ cm (f) 11 cm
(g) 27 cm (h) 177 cm

What progress have you made? (p 97)

1 (a) 60 cm (b) 80 cm (c) 1 m

2 (a) 40 cm (b) 35 cm
(c) 10 cm (d) 30 cm

Practice booklet

Section B (p 39)

1 (a) 0.2 m (b) 1.2 m (c) 2.8 m
(d) 4.4 m (e) 0.7 m

2 The pupil's sketches
(e) The child's head is below the windows.

Section C (p 40)

1 (a) 4 km or $4\frac{1}{2}$ km (b) $3\frac{1}{2}$ km
(c) $7\frac{1}{2}$ km

2 (a) Ingleborough Hill
(b) Pen-Y-Ghent

Section D (p 41)

1 (a) 50 m (b) 20 m
(c) 1000 m^2 (d) About 10 m
(e) About 40 m (f) About 50 m
(g) About 25 m

18 Questions and calculations

In this unit pupils decide which calculation is appropriate for a given word problem.

Practice booklet page 42

A Add, subtract, multiply or divide? (p 98)

T

'This generated good discussion about why some of the suggestions were wrong.'

◊ Some teachers have treated this entirely as an oral piece of work and practice in mental arithmetic. The main emphasis, however, should be on deciding what is the appropriate calculation for each question. All the correct calculations can be done mentally.

◊ You may need to point out that the arrows indicate which set of calculations go with which set of questions. Tell pupils that they will not need to use all the calculations.

B Your own questions (p 100)

T

◊ You could start by working orally with the group, asking for calculations for the questions about the crackers and for any questions you ask yourself.

◊ Give pupils time to think up some questions for themselves. Pupils could work in pairs or small groups on this. You could suggest that, at first, the questions are just about numbers of crackers, not involving money: 'How many crackers are in five boxes of gold crackers?' 'If a box of family crackers is shared between a family of four, how many crackers do they each get?' Concentrate for some time on generating questions, before eventually moving on to finding the operations which have to be done to answer them.

◊ When pupils have thought up plenty of simple questions, focus the discussion on the operations which need to be done to answer them. The main purpose of this exercise is to allow pupils to think up their own questions and reflect on and discuss the questions so they can see what operations are required to answer them.

B1 This question involves using money and it may be useful to get pupils to write the amounts simply in pence.

◊ Pupils may enjoy the opportunity to be creative in designing their own poster as a basis for questions. Some possible suggestions are:

'They loved this and produced some lovely posters.'

packs of icecreams and lollies	Christmas cards or gift tags
rides at a funfare	animals on a farm or at a zoo
bottles of drink	fish and chip shop
piles of coins	fruit and vegetables
packets of biscuits	newspapers

A Add, subtract, multiply or divide? (p 98)

A1 15 – 12 = 3

A2 15 + 4 = 19

A3 15 × 4 = 60

A4 12 ÷ 3 = 4

A5 12 – 3 = 9

A6 15 – 12 = 3

A7 16 + 10 = 26

A8 16 × 10 = 160

A9 16 ÷ 2 = 8

A10 24 – 6 = 18

A11 24 ÷ 6 = 4

A12 24 ÷ 4 = 6

A13 13 + 4 = 17

A14 13 × 3 = 39

A15 13 – 3 = 10

Practice booklet

Sections A and B (p 42)

1 (a) 10 × 6 = 60p
(b) 90 ÷ 10 = 9p
(c) 60 ÷ 12 = 5

2 (a) 15 – 6 = 9
(b) 48 ÷ 8 = 6p
(c) 8 + 5 = 13

3 (a) The pupil's question, calculation and answer on kiwi-fruit
(b) The pupil's question, calculation and answer on grapefruit

19 Spot the mistake 1

Pupils have to find common mistakes and correct them.
Most of the mistakes are based on actual pupils' work.

> *'This encourages explaining – which some of our students are not very good at.'*

◊ One way to organise this work is for the pupils to first work individually on a small number of questions. (Pupils may be less likely to forget what the questions are about if they work on only a small number at one time.) Then they can compare their results in pairs or small groups, discussing where they think the pupils have gone wrong. The teacher can go through the questions, drawing together the pupils' ideas.

For some, it will be enough that they can produce the correct solution to a problem. But encourage all pupils to try to explain orally or in writing why they think each mistake was made.

◊ Each answer includes an explanation of where the pupil has probably gone wrong. Pupils' answers may be less sophisticated.

Add, subtract, multiply and divide (p 101)

1 The pupil has written three thousand (3000) and then 48 to produce 300 048. But now the 3 is not in the thousands column.

The correct answer is 3048.

2 The 1 has not been carried into the 10s column but written down next to the 2.

The correct answer is 82.

3 The pupil has worked from left to right, but multiplication should be done before subtraction.

The correct answer is 4.

4 The pupil has turned the subtraction the wrong way round.

The correct answer is 12.

5 The calculator has done $8 \div 4$ first. The pupil should have used brackets or pressed '=' after '20 + 8'.

The correct answer is $(20 + 8) \div 4 = 7$.

6 In each column, the pupil has subtracted the smaller figure from the larger.

The correct answer is 46.

Handling data (p 101)

1 The highest frequency is 9 but this is not the mode.

The mode is the colour blue as it is the colour with the highest frequency.

2 The pupil has missed out the zero score and divided by 4 but there were 5 in the group.

The correct answer is $20 \div 5 = 4$.

3 There are three colours and one of them is red so the pupil has written $\frac{1}{3}$. However the yellow section is twice as big as the white and red sections.

The correct probability is $\frac{1}{4}$.

Fractions and decimals (p 102)

1 There are three pieces and one of them is shaded so the pupil has written $\frac{1}{3}$. But the pieces are not equal.

Only $\frac{1}{4}$ of the total area is shaded.

2 0.12 is 1 tenth and 2 hundredths. The pupil has thought of this as 12 tenths.

The largest number is 0.9.

3 The pupil has put the 3 in the tenths column but it should be under the 2 in the units column.

The correct answer is 15.6.

4 The answer to the calculation is correct but you cannot have 2.5 minibuses as you can't have half a minibus.

You would need 3 minibuses.

5 0.2 is two tenths but 0.03 is 3 hundredths and so on. The hundredths are smaller than the tenths.

The correct order is
0.03, 0.04, 0.06, 0.2, 0.5

Area (p 102)

1 The pupil has found the perimeter of the rectangle instead of the area.

The area is $4 \times 3 = 12\,\text{cm}^2$.

2 The pupil has multiplied all of the lengths but should have cut the shape into two rectangles.

The correct area is $21\ \text{cm}^2$.

20 Tens, hundreds and thousands review

This unit revisits place value and multiplication and division by 10, 100 and 1000. This is extended to decimals in unit 22 'Decimals and metric units'.

T	p 103 **A** Big numbers	Saying and writing large numbers
T	p 104 **B** Multiplying by 10, 100, 1000	Multiplying whole numbers by 10, 100, 1000
T	p 105 **C** Dividing by 10, 100, 1000	Dividing whole numbers by 10, 100, 1000
	p 106 **D** Multiply or divide?	Choosing whether to multiply or divide

Practice booklet pages 43 to 45

A Big numbers (p 103)

T

'Very good work for class discussion. Pupils enjoyed the quick-fire questions and boardwork'.

The activities in this section are designed to help pupils to write large written numbers in figures and vice versa. W1 Cheque it out in unit 1 'Number bites' is also useful practice in this skill. Simply reading out numbers in words and asking pupils to write these in figures is a good exercise.

◊ If pupils find this difficult you could write labelled columns on the board and write numbers in these columns. This will help pupils understand the place value of the numbers and is needed in later sections.

In 'Words and numbers' the pupils could also be asked to rearrange the numbers in order of size.

Big bingo

Read out the twelve numbers in a random order. If you find pupils avoiding the larger numbers, you could have some prepared cards.

B Multiplying by 10, 100, 1000 (p 104)

T

◊ Emphasise the movement of the digits as against 'adding 0s', as it will extend to decimals later. Zeros need to be seen simply as place holders to ensure other numbers are in the right column.

◊ Another simple bingo game is as follows:

Write four numbers on the board, for example 2, 4, 12, 20.
Each pupil makes a 2 by 2 grid and fills it with the multiples by 10, 100 or 1000 of any of the four given numbers, for example 20, 2000, 400, 1200 or 40, 400, 4000, 120.

You then call out one of the four numbers and choose randomly '× 10', '× 100' or '× 1000'. The winner is the first to tick off all of their numbers, or, say, two of their numbers.

T

C Dividing by 10, 100, 1000 (p 105)

◊ Again the emphasis must be on the movement of digits rather than losing 0s. Zeros disappear because they are no longer required as place holders.

D Multiply or divide (p 106)

B Multiplying by 10, 100, 1000 (p 104)

B1 (a) 350 (b) 30
(c) 1250 (d) 3500

B2 45 × 10 = 450
405 × 10 = 4050
4500 × 10 = 45 000
450 × 10 = 4500
4 × 10 = 40

B3 (a) 120 (b) 35 000 (c) 1200
(d) 1 000 000 (e) 1200 (f) 4000

B4 (a) 3300 (b) 24 000 (c) 1030
(d) 4000 (e) 306 000 (f) 21 000
(g) 25 100 (h) 7 002 000

B5 (a) 42 × 10 = **420**
(b) 52 × **100** = 5200
(c) 120 × **10** = 1200
(d) 89 × **1000** = 89 000
(e) **740** × 100 = 74 000
(f) **700** × 10 = 7000

C Dividing by 10, 100, 1000 (p 105)

C1 (a) 250 (b) 700 (c) 25
(d) 80 (e) 70 (f) 8

C2 (a) 600 (b) 570 (c) 24
(d) 6700 (e) 290 (f) 40
(g) 25 (h) 70 000

C3 (a) 2300 (b) 460 (c) 73 000
(d) 480 000 (e) 60 (f) 1020
(g) 70 000 (h) 240

C4 (a) 4 × 100 = **400**

(b) 5000 ÷ **1000** = 5

(c) 10 × **3** = 30

(d) 700 ÷ 100 = **7**

(e) **15** × 100 = 1500

(f) **45** × 10 = 450

(g) 14 × 100 = **14 000**

(h) 280 ÷ 10 = **28**

(i) 20 × 100 = **2000**

(j) 4200 ÷ **100** = 42

(k) 4000 ÷ 100 = **40**

(l) **30** × 100 = 3000

C5 (a)

30 —× 10→ **300** —÷ 100→ **3** —× 1000→ **3000**

(b)

70 —÷ 10→ 7 —× 1000→ **7000** —÷ 100→ **70**

(c)

43 —× 100→ **4300** —÷ 10→ **430** —× 1000→ **430 000**

(d)

7000 —÷ 10→ **700** —× 100→ **70000** —÷ 1000→ **70**

C6 In this order:
100 000 ÷ 1000 = 100
3700 ÷ 10 = 370
52 × 10 = 520
8500 ÷ 10 = 850
23 × 100 = 2300

***C7**

48 —× 100→ **4800** —÷ 10→ **480** —× 1000→ **480 000**

D Multiply or divide? (p 106)

D1 120 tins

D2 6 flats

D3 1500 tissues

D4 20 000 tickets

D5 45 000 donations

D6 350 000 stickers

D7 £30

D8 150 boxes

What progress have you made? (p 106)

1 (a) 40 050 (b) 203 006

2 (a) 35 000 (b) 420
(c) 460 000 (d) 3020

Practice booklet

Section A (p 43)

1 Fifty thousand four hundred

2 (a) It is 40 000 km round the Earth.

(b) It is 380 000 km to the Moon.

(c) The United Kingdom has an area of 244 000 square km.

(d) The Kalahari desert has an area of 260 000 square km.

(e) The English Channel has an area of 89 900 square km.

(f) Mount Everest is over 8800 metres high.

3 (a) Sixty thousand honey bees live in a colony.

(b) Three hundred thousand wood ants live in a wood ant nest.

(c) Two hundred and forty thousand tadpoles are in a pond.

(d) A swan has over twenty-five thousand feathers.

(e) Brachiosaurus weighed seventy-three thousand kg.

(f) Tyrannosaurus lived sixty-five million years ago.

Sections B and C (p 44)

1 (a)

×	10	100	1000
63	630	**6300**	**63000**
40	**400**	4000	**40000**
28	**280**	**2800**	**28000**

(b)

×	10	**100**	1000
7	**70**	700	**7000**
58	580	**5800**	**58000**
12	**120**	**1200**	12000

2 (a) 59 × 100 = **5900**
(b) 1070 ÷ **10** = 107
(c) 72 × **10** = 720
(d) **88 000** ÷ 100 = 880
(e) 463 × 1000 = **463 000**
(f) **72** × 100 = 7200
(g) 630 ÷ **10** = 63
(h) 610 × 100 = **61 000**
(i) **620** ÷ 10 = 62
(j) **550** × 10 = 5500
(k) 62 000 ÷ **1000** = 62
(l) 91 × 1000 = **91 000**
(m) 50 700 ÷ 100 = **507**
(n) 9000 ÷ **10** = 900

3 A and D: 680 B and J: 68 000
C and H: 680 000 E and F: 6800
G and I: 68

4 (a)
50 —× 100→ **5000** —÷ 10→ **500** —× 1000→ **500 000**

(b)
75 —× 10→ **750** —× 100→ **75000** —÷ 1000→ **75**

Section D (p 45)

1 £80 000

2 £45 000

3 18

4 £50

5 36 000

6 1200

7 520

8 6000

9 250

21 Solids

T T T

p 107 **A** Describing	Practical exercise to practise the language of three dimensions
p 107 **B** Drawing 3-D objects	Using triangular dotty paper to draw 3-D objects
p 110 **C** Cuboids	Discussion of the definition of a cuboid Investigation of different cuboids made from the same number of cubes
p 111 **D** Counting cubes	Finding the number of cubes in a cuboid or simple prism

Essential	**Optional**
Multilink cubes (about 6 per pupil) Triangular dotty paper Sheet 241 **Practice booklet** pages 46 to 47	Collection of cuboids, prisms and other objects

A Describing (p 107)

T

The purpose of this activity is to encourage pupils to be precise when describing three-dimensional shapes.

Multilink cubes (about 6 per pupil)

'They enjoyed this, but I had to push them to use correct language.'

◊ You could start by giving an oral description of a shape (which the pupils cannot see) made from multilink, and asking each pupil to make the shape. Compare the results and discuss how any mistakes arose.

Then, in pairs, one pupil makes a solid with a small number of cubes and keeps it concealed. The other pupil tries to make the same shape. This could be done by the first pupil describing their shape or by the second pupil asking questions to which only 'yes' or 'no' can be answered.

◊ The class could make a list of words they have found useful, such as 'layer', 'face', 'corner', 'behind'. They should know that we often write 3-D to stand for 'three dimensions' or 'three-dimensional'.

B Drawing 3-D objects (p 107)

The main objective of this exercise is to allow pupils to draw 3-D objects in true isometric projection using triangular dotty paper. Weaker pupils will find this difficult and may need lots of practice.

Multilink cubes, triangular dotty paper, sheet 241

T

◊ Pupils will need to be shown how to orientate the triangular dotty paper correctly – this is a vital first step! There need to be obvious columns (rather than rows) of dots.

◊ The drawing of the L shape could be demonstrated on the board or with an OHP transparency of dotty paper. One method of drawing is to draw the plane letter first and then draw in lines to give thickness.

*'I challenged them to draw their own names. The pupils were **very** enthusiastic (an unusual event!).'*

◊ As a group activity pupils could be asked to draw all the possible objects using exactly 4 multilink cubes. There are 8 altogether. Discuss, as necessary, where shapes are the same but in a different orientation.

B1 As an extension you could ask pupils to draw their own initials, or their names.

C Cuboids (p 110)

Optional: A variety of everyday objects some of which are cuboids, some not; multilink cubes

T

◊ A collection of objects, some of which are not cuboids, is useful.
It is useful to have to hand objects which would fit wrong definitions – for example if someone says 'a flat box' have a flattened carton.
'All rectangular faces' is actually an indisputable definition!

C1, C2 Pupils may find multilink cubes helpful.

C3 This could be linked to work on factors. Pupils could be asked why there is only one way to make a cuboid using a prime number of cubes.

D Counting cubes (p 111)

In this section pupils should be encouraged to see that they can find the number of cubes in a cuboid or prism by finding the number in one layer, and then multiplying by the number of layers.

B Drawing 3-D objects (p 107)

B1 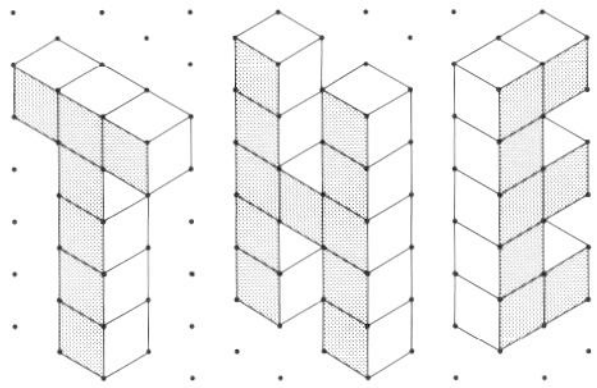

B2 (a) 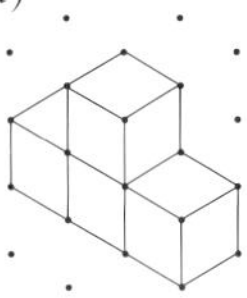 (b) 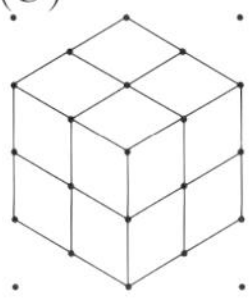 (c) 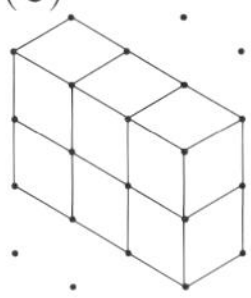

B3 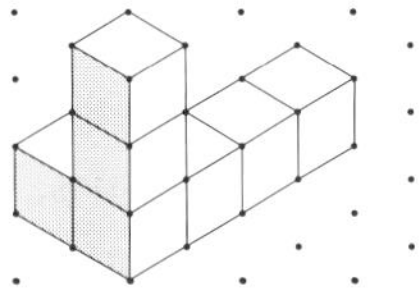

B4 (a) 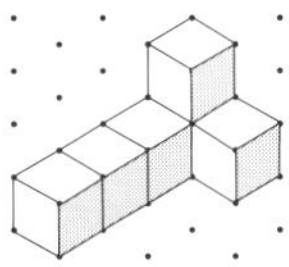 (b) 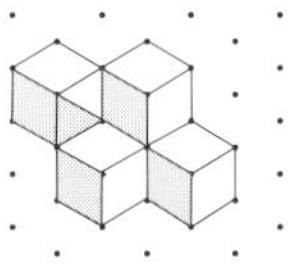

(c) 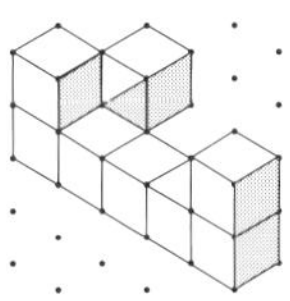 (d) 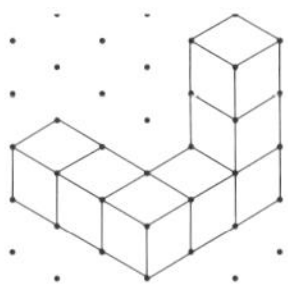

B5 C is correct.

C Cuboids (p 110)

C1 3 by 4 by 1 1 by 12 by 1 2 by 3 by 2

C2 1 by 1 by 8 1 by 2 by 4 2 by 2 by 2

C3 (a) 4 cuboids (b) 6 cuboids

D Counting cubes (p 111)

D1 (e) is biggest. (f) is smallest.
In order f, d, (a and b), c, e

D2 (a) 24 (b) 28 (c) 27 (d) 80

D3 (a) 10 (b) 24 (c) 30 (d) 15

What progress have you made? (p 112)

1 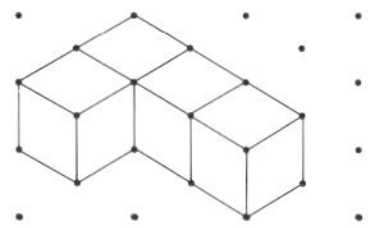

2 The pupil's 3 cuboid-shaped objects

3 A and C

4 (a) 12 (b) 30

5 (a) 16 (b) 20

Practice booklet

Section B (p 46)

1 (a) 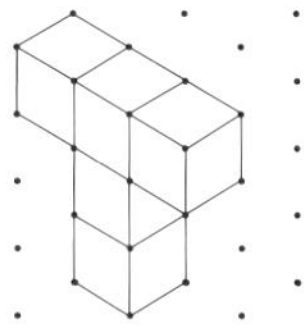 (b) 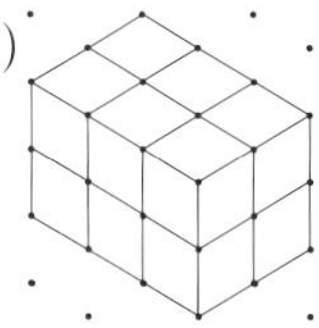

(c) 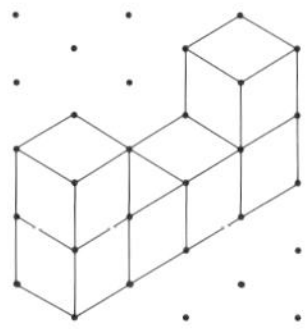

2 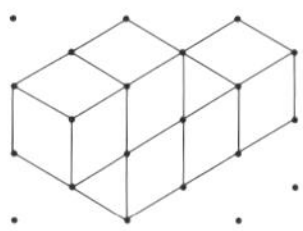

Sections C and D (p 47)

1 The pupil's two different sketches of cuboids made from 18 cubes

2 (a) 36 (b) 40 (c) 32 (d) 36

3 (a) 12 (b) 40 (c) 18

Review 3 (p 113)

1 (a) 151 g (b) 146.5 g
(c) 157 g (d) the males

2 (a) 300 km (b) 500 km
(c) 1300 km (d) 1100 km

3 (a) 2800 (b) 4000 km

4 (a) 24×8 (b) $24 \div 8$
(c) $24 - 8$ (d) $24 + 8$

5 (a) The pupil has taken 7 from 100, or not broken down 100 correctly into 90 and 10.

The correct answer is 83.

(b) The pupil has not arranged the numbers in order of size before finding the middle one.

The correct answer is 3.

(c) The pupil has read each division on the scale as 0.1, but it is 0.01.

The correct answer is 3.99.

(d) The rectangle has four sides and the pupil has only added up two.

The correct answer is 12 cm.

6 (a) 300 000 fans greet team
(b) 2030 breakdowns on railways
(c) 1 000 001 ways to win!
(d) 300 207 umbrellas lost

7 (a) 2200 (b) 430 (c) 41 000
(d) 500 (e) 4500 (f) 20 100

8 (a) £20 000 (b) £300
(c) 2000 (d) 5

9 (a) 18 (b) 48 (c) 12 (d) 20

10 (a) and (b) are cuboids.

Mixed questions 3 (Practice booklet p 48)

1 (a) 55 (b) 11 (c) 13.5

2 (a) 12 m (b) 3 m by 2 m
(c) 3 m (d) $128\ m^2$

3 (a) $40 \div 10$; 4p
(b) $40 + 10$; 50 stickers
(c) 40×10; 400p or £4.00
(d) $40 - 10$; 30 stickers

4 A and S, B and T, C and R, D and Q, E and P

5 (a) 25 000 (b) 3200
(c) 40 000 (d) 8
(e) 1600 (f) 790

6 (a) 12 cubes (b) 14 cubes
(c) 12 cubes

22 Decimals and metric units

This unit extends multiplication and division by 10, 100 and 1000 to decimals and links this to changing between metric units.

Essential

Sheet 242

Practice booklet pages 50 to 52

A Multiplying decimals by 10 (p 115)

The purpose of the initial activity is to establish that $0.7 \times 10 = 7$ from a purely practical perspective. You could also ask how long a row of 100 nuts would be.

B Multiplying and dividing decimals by 10 and 100 (p 116)

'A neat idea.'

'Very useful – several students used their grid regularly.'

Using tracing paper and a grid will help pupils to appreciate that it is the digits that move in relation to the decimal point rather than the other way round. This can be demonstrated with an OHP transparency. Draw a grid with a decimal point, like this: 

Put tracing paper over the grid and write the number to be multiplied. Then move the tracing one place to the left.

Obviously, multiplication by 100 and division can be shown in a similar way. This is also a good opportunity to discuss where extra zeros are required or can be dropped.

C Metric units (p 118)

In this section multiplication and division by 10 and 100 are linked to changing between centimetres/millimetres and metres/centimetres.

D Multiplying and dividing decimals by 1000 (p 120)

The bingo game described in unit 19 'Tens, hundreds and thousands review' on page 103 is easily extended to multiplying decimals by 10, 100 or 1000, using starting numbers such as 0.2, 0.4, 0.06, 0.05. It is probably best to keep to multiplying by 10 and 100 at first. A coin labelled 10 and 100 on opposite faces can be flipped to decide what to multiply by. Pupils then fill their grids with numbers such as 2, 20, 4, 40, 0.6, 6, 0.5 or 5.

You may need to write the list of possible grid numbers on the board at first, until pupils have understood.

E Changing between units (p 121)

Oral questions (p 122)

T

Examples of questions you could ask about the pictures:

1 What is the diameter in centimetres of the 1p coin?

2 What is the diameter in centimetres of the 2p coin?

3 If five 1p coins are laid in a line, what is the total length in centimetres?

4 How many 1p coins would make a total length of 20 centimetres?

5 What is the total weight of the three boxes of corn flakes, in grams?

6 What is the total weight in kilograms?

7 How many of the 250 g boxes would weigh 1 kilogram altogether?

8 What is the weight in kilograms of three of the 500 g boxes?

9 What is the total amount of milk, in litres, in the first three cartons?

10 What is the total amount in the last two containers?

11 How much milk is there altogether?

12 How many 500 ml cartons are needed to fill up a 1.5 litre container?

F Tens, hundreds and thousands review (p 122)

Sheet 242

A Multiplying decimals by 10 (p 115)

A1 8 cm

A2 (a) 7 (b) 9 (c) 2 (d) 1

A3 13 cm

A4 (a) 15 cm (b) 18 cm (c) 25 cm
(d) 30 cm (e) 125 cm

A5 (a) 17 (b) 19 (c) 52
(d) 105 (e) 254

B Multiplying and dividing decimals by 10 and 100 (p 116)

B1 (a) 59 (b) 234 (c) 3
(d) 86 (e) 104

B2 (a) 14.5 (b) 276.2 (c) 7.4
(d) 126.3 (e) 50.4

B3 (a) 6.5 (b) 40.6 (c) 0.8
(d) 5 (e) 70.2 (f) 1.33
(g) 0.64 (h) 10.38 (i) 0.4
(j) 6.05

B4 (a) 5.4 (b) 42 (c) 146
(d) 1.46 (e) 30.7 (f) 1.382
(g) 6.7 (h) 24

B5 (a) 704 (b) 0.345 (c) 97
(d) 0.518 (e) 60 (f) 108
(g) 0.32 (h) 3

B6 (a) 10 × 6.3 = **63**
(b) 3.1 × **100** = 310
(c) 420 **÷** 10 = 42
(d) 0.89 × 10 = **8.9**
(e) 6.07 × **100** = 607
(f) 0.03 × **100** = 3
(g) 4 **÷ 100** = 0.04
(h) 0.86 **× 100** = 86
(i) **40.2** ÷ 100 = 0.402

B7 (a) CHEESE (b) BREAD
(c) PIZZA (d) SOUP

B8 (a) 6 → × 10 → 60 → ÷ 100 → 0.6 → ÷ 10 → 0.06
(b) 0.8 → ÷ 10 → 0.08 → × 100 → 8 → ÷ 10 → 0.8
(c) 0.2 → × 10 → 2 → × 100 → 200 → ÷ 10 → 20
(d) 3 → ÷ 100 → 0.03 → × 10 → 0.3 → × 100 → 30

B9 In order of size
0.02 × 100 = 2
0.42 × 10 = 4.2
47 ÷ 10 = 4.7
0.6 × 10 = 6
650 ÷ 100 = 6.5

***B10** 3.7 → × 10 → 37 → ÷ 100 → 0.37 → ÷ 10 → 0.037

B11 7 litres

B12 0.072 g

C Metric units (p 118)

C1 (a) 5 cm = 50 mm
(b) 7.3 cm = 73 mm
(c) 3.1 cm = 31 mm
(d) 6.3 cm = 63 mm
(e) 1.6 cm = 16 mm

C2 (a) 74 mm (b) 115 mm
(c) 70 mm (d) 200 mm
(e) 6 mm

C3 (a) 4 cm (b) 4.2 cm (c) 13.8 cm
(d) 50 cm (e) 0.9 cm

C4 (a) 86 (b) 13.9 (c) 820

C5 In order: Aaron, Shari, Lizzie, Martin, Sanjay

C6 (a) 300 cm (b) 250 cm (c) 50 cm
(d) 75 cm (e) 8 cm

C7 (a) 6 m (b) 1.4 m (c) 0.35 m
(d) 0.7(0) m (e) 0.09 m

C8 (a) 205 cm (b) 1.07 m (c) 0.68 m
(d) 2 cm

C9 (a) 1.25 (b) 0.38 (c) 120
(d) 25 (e) 30 (f) 0.09

D Multiplying and dividing decimals by 1000 (p 120)

D1 (a) 4000 m (b) 3700 m (c) 400 m
(d) 20 000 m (e) 80 m

D2 (a) 7 km (b) 6.5 km (c) 0.25 km
(d) 0.4 km (e) 0.02 km

D3 (a) 3450 (b) 0.27 (c) 31 300
(d) 60

D4 (a) 1200 g (b) 1.3 kg (c) 0.65 kg
(d) 2350 g

D5 (a) 3.3 kg (b) 0.425 kg (c) 0.785 kg
(d) 0.075 kg

D6 (a) 2500 g (b) 280 g

E Changing between units (p 121)

E1 2.25 litres

E2 4500 metres

E3 1.5 kilometres

E4 20 000 metres

E5 125 centimetres

E6 0.65 metre

E7 108 centimetres

E8 1.125 kilograms

E9 (a) 5300 g (b) 0.85 kg
(c) 1.4 litres (d) 3750 millilitres

F Tens, hundreds and thousands review (p 122)

F1 (a)

×	0.5	0.3	0.6
10	5	3	6
100	50	30	60
1000	500	300	600

(b)

×	1.5	3.5	3.2
100	150	350	320
1000	1500	3500	3200
10	15	35	32

(c)

×	0.09	0.04	0.06
100	9	4	6
10	0.9	0.4	0.6
1000	90	40	60

(d)

×	0.51	0.35	0.52
10	5.1	3.5	5.2
100	51	35	52
1000	510	350	520

(e)

×	0.4	8	1.6
10	4	80	16
100	40	800	160
1000	400	8000	1600

(f)

×	12	6.5	0.25
100	1200	650	25
1000	12000	6500	250
10	120	65	2.5

F2 (a) 30 × 10 = **300**
300 ÷ 100 = **3**
3 ÷ 10 = **0.3**
0.3 ÷ 10 = **0.03**

(b) 0.05 × 10 = **0.5**
0.5 × 100 = **50**
50 ÷ 10 = **5**
5 ÷ 10 = **0.5**

F3 (a) $7 \div \mathbf{10} = 0.7$
$0.7 \times \mathbf{100} = 70$
$70 \times \mathbf{100} = 7000$
$7000 \div \mathbf{1000} = 7$

(b) $20 \times \mathbf{100} = 2000$
$2000 \div \mathbf{10} = 200$
$200 \div \mathbf{100} = 2$
$2 \div \mathbf{10} = 0.2$

What progress have you made? (p 122)

1 (a) 75 (b) 345 (c) 6.53
(d) 3270 (e) 3.28 (f) 0.575

2 (a) 5.06 m (b) 1400 ml
(c) 5.67 km (d) 3250 g

Practice booklet

Sections A and B (p 50)

1 (a) 42 (b) 136 (c) 31.5
(d) 604 (e) 258.3 (f) 80.7

2 (a) 5.8 (b) 24.7 (c) 30.6
(d) 8.26 (e) 6.03 (f) 0.18

3 (a) 500 (b) 50 (c) 5
(d) 0.05 (e) 50 (f) 5000
(g) 5 (h) 500 (i) 5000

4 (a) COBRA (b) HORSE
(c) MOOSE (d) SHEEP

Section C (p 51)

1 (a) 2.7 cm (b) 35 mm
(c) 30 mm (d) 5 cm

2 (a) 90 cm (b) 500 mm
(c) 12.5 cm (d) 6.2 cm

3 (a) 5.3 cm (b) 60 mm (c) 250 mm
(d) 10.6 cm (e) 0.8 cm (f) 143 mm

4 (a) 1.75 m (b) 550 cm (c) 3.15 m
(d) 240 cm (e) 270 cm

5 (a) 500 cm (b) 280 cm (c) 8 m
(d) 1.53 m (e) 86 cm (f) 4 cm
(g) 0.89 m (h) 1200 cm (i) 0.07 m

Sections D and E (p 52)

1 (a) 0.1 km (b) 0.2 km (c) 0.4 km
(d) 0.8 km (e) 1.5 km (f) 3 km
(g) 5 km (h) 10 km

2 (a) 11 200 m (b) 96 000 m
(c) 20 000 m (d) 64 000 m

3 (a) 46 500 (b) 543 (c) 360
(d) 3800 (e) 62.8 (f) 0.64

4 $7 \times 1000 = 7000$ $0.7 \times 1000 = 700$
$70 \times 1000 = 70\ 000$

5 (a) 3800 m (b) 4.5 litres
(c) 0.6 kg (d) 1.38 m
(e) 5.4 km (f) 700 millilitres

6 0.6 kg

7 2500 millilitres

23 Estimation and units

Essential	**Optional**
Chalk, tape measures	Garden stick markers
Trundle wheel	Local map
Sheet 243	1 kg bag of sugar, 2 litre plastic bottle, etc
Practice booklet pages 53 to 54	

A Good judgement (p 123)

T

The purpose of this section is to provide pupils with 'handles' on which they can base their estimates. Pupils may have some useful ones already from their own experiences; for example, swimmers may have a good idea what 25 m is. Pupils should be encouraged to use these and describe them to others.

Each part asks pupils to guess a distance, confirm by measurement and then use this to estimate other distances.

> Chalk, tape measure, trundle wheel
> Optional: garden stick markers, local maps

'We had a lot of fun with this. We collected the estimates of the whole class and calculated the means.'

◊ **How long is 1 m?** Drapers sometimes estimate 1 m by stretching their arms. Large sheets of paper (e.g. cheap wall-lining paper) can be Blu-tacked to the wall for this activity or chalk against a suitable outdoor wall.

◊ **How long is 100 m?** Pupils could mark their guesses with plant markers which are very cheap and reusable. After one guess at 100 m, pupils could then try again and see if guesses improve.
Only one trundle wheel is needed – it is useful if the class follow round whoever is measuring and watch their guesses being confirmed or otherwise. Tape measures could be used but are clumsy.

◊ **How far is a mile?** This activity is enhanced by having a prepared list of local key places which pupils can estimate the distance from the school.

A4 A hand is officially 4 inches. Pupils interested in horses may know that this is about 10 cm. Note that when a horse is described as 15.2 hh this means 15 hands and 2 inches.

B Converting (p 125)

T

Pupils need to learn the metric equivalents of the imperial units mentioned in this section. It will be useful to return to these repeatedly to reinforce them, for example while giving oral questions.

Optional: 1 kg bag of sugar, 2 litre soft drink bottle and other packets and containers marked with their metric weight or capacity

You may find it useful to bring in examples of objects of known weight or capacity – bags of sugar or soft drink bottles etc so that you can discuss their rough equivalents in imperial units. Then go on to discuss how to change from (say) 4 pounds to kilograms.

The questions are designed to be answered without a calculator. Most involve simple doubling (perhaps more than once), and pupils should be able to do the arithmetic mentally.

◊ Pupils' heights and weights may give opportunity for conversion between inches and centimetres, and between pounds and kilograms. These will involve more complicated calculations than are expected in the questions.

◊ Pupils could use a spreadsheet to produce ready reckoners for converting between metric and imperial measurements.

C Graphs (p 127)

Sheet 243

A Good judgement (p 123)

A1 The pupil's estimate

A2 The pupil's estimate (a chair is about 40 cm wide)

A3 The pupil's estimate (a typical car park space is about 2 m wide)

A4 (a) 60 in = 150 cm

(b) 68 in = 170 cm

(c) 64 in = 160 cm

A5 The pupil's estimate (the Skylon Tower is in fact 555 m tall)

Handy measurements

A cubit is generally considered to be about 50 cm long, giving the Ark a length of about 150 m.

B Converting (p 125)

B1 (a) 5 cm (b) 10 cm

(c) 15 cm (d) $7\frac{1}{2}$ cm

B2

B3 (a) 4 m by 2 m by 3 m
(b) About £40

B4 60 grams butter
120 grams sugar
240 grams flour
15 grams baking powder

B5 (a) 2 kg (b) 4 kg (c) 5 kg (d) 10 kg

B6 (a) 2 litres (b) 4 litres
(c) 5 litres (d) $2\frac{1}{2}$ litres

B7 (a) 18 litres (b) 27 litres
(c) 36 litres

B8 The Mini holds about 27 litres, so it holds less.

B9 (a) Potato and leek soup (serves six)
1 kg white leeks
$\frac{1}{2}$ **kg** potatoes
1 litre chicken stock
$\frac{1}{4}$ **litre** cream
15 grams chopped parsley
salt and pepper to taste

Trim, wash and chop the leeks into **$2\frac{1}{2}$ cm** pieces...

(b) About $4\frac{1}{2}$ kg (c) About 35 grams

B10 With the pupil's explanations
(a) Yes (b) Yes (c) Yes
(d) No (e) Yes

C Graphs (p 127)

C1 (a) The pupil's check
(b) The pupil's check

C2 (a) 4 kg (b) 15 kg
(c) 23 kg (d) 5 kg

C3 (a) 20 pounds (b) 46 pounds
(c) 55 pounds (d) 31 pounds

C4 Chloe weighed 4 kg = 9 pounds and her mother 10 pounds = 4.5 kg, so her mother weighed more.

C5 About 7 kg

C6 56 pounds = 25.5 kg, 20 kg = 44 pounds, so 56 pounds is heavier
Difference in weight = 5.5 kg

C7 (a) Lager sales

Tuesday	11 to 11.5 litres
Wednesday	24.5 to 25 litres
Thursday	30 litres
Friday	39 litres

(b) 80 pints = 45 to 45.5 litres, so they sell about 24.5 to 25 litres more of red wine.

(c) About 34 litres

(d) 15 pints of lemonade = about 8.5 litres, so he should have enough.

What progress have you made? (p 128)

1 2 metres

2 25 cm

3 (a) About 30 g (b) About 3 feet
(c) About 8 km (d) About $4\frac{1}{2}$ litres

4 (a) 90 g (b) 24 km
(c) 5 cm (d) 5 litres

5 (a) 4 gallons = 18 litres, so an empty tank will hold 15 litres
(b) 5 gallons = $22\frac{1}{2}$ litres so 20 litres will not fill the tank

6 (a) 6.5 to 7 gallons (b) 9 litres
(c) 29 to 29.5 litres

Practice booklet

Sections A and B (p 53)

1 (a) The pupil's classroom length and width
(b) The pupil's classroom area
(c) Answer (b) × £15

2 (a) centimetres (b) metre
(c) kilogram (d) litre

3 (a) 3 m wide by 5 m long by 4 m high

(b) The area is about 64 m^2, so 13 tins are needed at a cost of £104.

4 (a)

Lytefoe	8 km
Heal	16 km
Lanche	24 km

(b) 40 km

5

Vespa	$4\frac{1}{2}$ litres
Tiger cub	9 litres
Land Rover	45 litres
Dennis	90 litres
Bristol bus	225 litres

Sections C (p 54)

1 (a) 9 metres (b) $4\frac{1}{2}$ metres
(c) 16 metres (d) 23 metres

2 (a) 10 feet (b) 65 to 66 metres
(c) 39 to 40 feet (d) 69 feet

3 (a) $3\frac{1}{2}$ metres (b) 18 metres
(c) 15 metres

4 (a) The 1996 jump by about 2 metres.

(b) Randy Barnes, 1996 put was furthest by about 10 metres.

24 Spot the mistake 2

As in 'Spot the mistake 1' pupils have to find common mistakes and correct them.

Measures (p 129)

1 The 20 has been added to the 45 and written as 65 minutes.

The correct answer is 15:05.

2 The pupil thinks there are only 100 grams in a kilogram instead of 1000 grams.

The correct answer is $1000 - 20 = 980$ grams.

3 The time 3:00 has been treated as 300 and, as $300 - 50 = 250$, the time has been given as 2:50.

The correct answer is 2:10.

4 The pupil thinks that any number after the decimal point is centimetres. But 4.7 metres is equivalent to 4 metres and 70 centimetres.

The correct answer is 4.07 metres.

5 The pupil has divided by 100 instead of 1000.

The correct answer is $2000 \div 1000 = 2$ (km).

Coordinates (p 129)

1 The coordinates have been read the wrong way round.

The correct coordinates are (3, 0).

2 The pupil has not used equally spaced scales on the axes.

The scales should be labelled 0, 1, 2, 3, 4, ... on both axes and the points correctly plotted.

Angles (p 130)

1 The pupil has gone the wrong way from a multiple of 10, going 4 degrees up from 50 to obtain 54°.

The correct answer is 46°.

2 The outer scale on the protractor has been read to give 60° but the angle is obtuse.

The inner scale gives the correct answer of 120°.

Symmetry (p 130)

1 Although the diagonals cut the rectangle in half, neither are lines of symmetry.

Only the horizontal and vertical lines are lines of symmetry.

2 Although the diagonals cut the parallelogram in half, neither are lines of symmetry.

This parallelogram has no lines of symmetry.

Money (p 130)

1 The division is the wrong way round.

The correct answer is $5 \div 10 = 0.5$ so each packet cost £0.50 or 50p.

2 The pupil thinks that any number after the decimal point is pence. An extra 0 needs adding so £1.3 is equivalent to £1.30.

The correct answer is £1.30.

3 The calculation is correct but 5.2 needs a 0 adding to make £5.20.

The bars cost £5.20.

25 Simple substitution

p 131	**A** Review	Review of *Book T1*, 'Using rules' and 'Calculate in order'.
T p 133	**B** The right order	Evaluating expressions with brackets, and where you cannot work from left to right.
T p 136	**C** Solving equations	Solving simple equations
p 137	**D** Words to symbols	Translating situations into algebra

Essential

Sheet 244
Normal dice or dice marked 0, 0, 1, 1, 2, 2, and 4, 4, 5, 5, 6, 6 (a pair per group)
Counters

Practice booklet pages 55 to 57

A Review (p 131)

This material should be familiar to the students but some may need extra help, especially in A6 and beyond where the problems are more abstract.

B The right order (p 133)

Sheet 244, dice and counters

T

Unit 23 in *Book T1*, 'Calculate in order' introduced the rules for evaluating an expression in a numerical context. Here we extend this to algebraic expressions. You will want to discuss several examples with pupils to bring out the various pitfalls that can arise.

'The pupils devised their own games and tried them out on the rest of the class.'

Grand Prix

This simple game will give pupils plenty of practice in evaluating expressions in a situation where feedback is immediate. It can be played with a normal 1 to 6 dice, but will promote better understanding if pupils use the 0–2 and 4–6 dice. Pupils then choose at each turn which of the two dice they wish to use.
An alternative way to end could be that you can only finish by getting the exact number to land on the 'start' square again.
Book S2 sheet 215 uses a version of this game with harder expressions.

C Solving equations (p 136)

T

This section introduces the words 'equation' and 'solve'. It is intended that pupils treat solving equations as a small step from number puzzles, and the teacher-led introduction may be used for this. It is not intended that the idea of balancing should be introduced; most pupils will spot the answer or use trial and error.

C5 It would be simple to produce a set of 'Equation snap' cards for pupils to play.

D Words to symbols (p 137)

We would not expect all pupils using the T materials to be ready for this section. It may be omitted without causing problems later.

A Review (p 131)

A1 (a) 15 (b) 50 (c) 16
(d) 3 (e) 12 (f) 5

A2 (a) £17 (b) £25
(c) $c = 2 \times n + 5$

A3 Polisher $c = 3n + 2$
Jack $c = 2n + 3$
Heater $c = 4n + 3$
Drill $c = 3n + 4$

A4 (a) £21 (b) £36
(c) **£5** per day plus an extra **£6**

A5 (a) 35 (b) 65 (c) £17 (d) 5 days

A6 (a) 24 (b) 17 (c) 3
(d) 9 (e) 2 (f) 60

A7 (a) 20 (b) 22 (c) 14 (d) 12

A8 (a) 28 (b) 18 (c) 208 (d) 16

A9 (a) 6 (b) 11 (c) 5 (d) 51

A10 A gives the biggest value (34).

A11 D gives the smallest value (2).

B The right order (p 133)

B1 Pupils should include the working.
(a) 20 (b) 14 (c) 12 (d) 12

B2 (a) 9 (b) 24 (c) 3
(d) 20 (e) 7 (f) 6

B3 IT IS SOLVED

B4 (a) 60 plates (b) 100 plates
(c) B: $p = 2(n + 5)$

B5 (a) 90 straws (b) 150 straws
(c) D: $s = 5n - 10$

B6 A: $g = 3(n + 15)$

B7 160 balloons, 56 tables, 450 poppers

B8 (a) 68 (b) 48 (c) 60 (d) 96

B9 (a) 10 (b) 7 (c) 2 (d) 12

***B10** (a) 9 (b) 25 (c) 1 (d) 16

***B11** (a) 25 (b) 27 (c) 5
(d) 16 (e) 19 (f) 26

C Solving equations (p 136)

C1 (a) 3 (b) 4 (c) 8
(d) 6 (e) 12 (f) 5

C2 (a) $n = 5$ (b) $j = 10$ (c) $f = 8$
(d) $t = 5$ (e) $h = 2$ (f) $g = 5$

C3 (a) $n = 5$ (b) $b = 5$ (c) $c = 7$
(d) $n = 10$ (e) $a = 12$ (f) $m = 12$

C4 (a) $n = 4$ (b) $f = 12$ (c) $f = 2$
(d) $n = 10$ (e) $b = 5$ (f) $y = 22$

C5 A and D ($n = 7$); B and F ($n = 6$)
C and E ($n = 5$)

***C6** (a) $z = 7$ (b) $w = 8$ (c) $m = 6$

D Words to symbols (p 137)

D1 (a) $2n$ and $n + n$
(b) $n + 2$ and $2 + n$

D2 (a) $3n$ (b) $n - 3$ (c) $3n - 3$

D3 (a) $4p$ (b) $2p + 12$

D4 (a) $3n$ (b) $n - 20$

What progress have you made? (p 139)

1 (a) 6 (b) 24 (c) 3 (d) 0

2 (a) 5 (b) 3 (c) 9 (d) $3\frac{1}{2}$

3 (a) 10 (b) 28 (c) 20 (d) 15

4 $n + 3$ or $3 + n$

5 (a) $3s$ (b) $s - 50$

Practice booklet

Section A (p 55)

1 (a) £13 (b) $t = 3g + 1$

2 (a) £7 (b) Sally can play 4 games.
(c) £4.50 (d) Rob can play 6 games.

Section B (p 55)

1 (a) When $n = 3$
$3n + 2 = 11$ $3(n + 2) = 15$
$6 + n = 9$ $3(n - 1) = 6$
(b) The card with the biggest value is $3(n + 2)$.

2 (a) When $p = 5$
$10 + 4p = 30$ $10(p - 1) = 40$
$10(2 + p) = 70$ $10p - 2 = 48$
(b) When $p = 3$
$10 + 4p = 22$ $10(p - 1) = 20$
$10(2 + p) = 50$ $10p - 2 = 28$
(c) When $p = 10$
$10 + 4p = 50$ $10(p - 1) = 90$
$10(2 + p) = 120$ $10p - 2 = 98$

3 (a) When $s = 6$
$\frac{s}{2} + 2 = 5$ $\frac{s}{3} + 3 = 5$
$10 + \frac{s}{3} = 12$
(b) When $s = 12$
$\frac{s}{2} + 2 = 8$ $\frac{s}{3} + 3 = 7$
$10 + \frac{s}{3} = 14$

Section C (p 56)

1 (a) 7 (b) 6 (c) 11
(d) 10 (e) 20 (f) 36

2 (a) 3 (b) 10 (c) 6
(d) 6 (e) 9 (f) 101

3 (a) 3 (b) 4 (c) 8
(d) 27 (e) 2 (f) 32

4 (a) 6 (b) 6 (c) 5
(d) 18 (e) 24 (f) 5

5 (a) 2 (b) 2 (c) 10
(d) 20 (e) 25 (f) 9

Section D (p 57)

1 (a) Colin has $w - 3$ chocolate bars.
(b) Ella has $2w$ chocolate bars.
(c) (i) Alex has $w + 5$ bars.
(ii) Colin has $w - 3 + 5$ bars $= w + 2$ bars.
(iii) Ella has $2w + 5$ bars.

2 (a) A plate costs $c + 10$ pence.
(b) 6 cups cost $6c$ pence.
(c) A plate and a cup cost
$2c + 10$ or $10 + 2c$ or $c + c + 10$.
(d) The cost of two plates is $2c + 20$ pence.

26 Multiplication

The unit revises multiplications of single digit by two-digit numbers and multiples of 10. It introduces multiplying a two-digit number by two- and three-digit numbers.

Practice booklet pages 58 to 60

A Review (p 140)

This section revises the work in *Book N* unit 5, 'Multiplication'.

B More multiplication (p 141)

T

◊ The following points should be brought out in the initial discussion.

- Order doesn't matter for multiplication: $2 \times 3 \times 4 = 4 \times 2 \times 3$.
- Multiplying a number which is a multiple of 10 can be carried out by writing the number as a simple number multiplied by 10, 100 etc, so $30 \times 20 = 3 \times 10 \times 2 \times 10 = 3 \times 2 \times 10 \times 10 = 6 \times 100 = 600$.

The second example deals with the case where an extra zero is produced by $4 \times 5 = 20$.

◊ This method (rather than 3×2 followed by adding zeros) can be adapted for decimals at a later stage (e.g. $3.6 \times 20 = 3.6 \times 2 \times 10 = 7.2 \times 10 = 72$).

C Area and multiplication (p 142)

T

◊ The teacher-led discussion should establish that multiplications can be split into simpler multiplications and that this process can be recorded in a table. The illustrations at the beginning should provide pupils with a visual reminder of the process.

D Tables to multiply (p 144)

T

◊ The teacher-led discussion should establish the same points as in section C.

Four digits

◊ As there are 12 different multiplications (41 × 32 being the same as 32 × 41), it may help if pupils work in groups. After identifying the different multiplications, pupils could share the task of finding the results.

◊ For the digits 1, 2, 3, 4, the largest result is 41 × 32 = 1312 and the smallest is 13 × 24 = 312.

For the largest result, the largest numbers 4 and the 3 are put in the tens place. The choice is then between 42 × 31 and 41 × 32. Think about the units × tens products. In the second case the larger units figure multiplies the larger tens figure, so this gives the larger result. A similar argument can be used for the smallest result.
Using the same arguments the largest result with 1, 2, 3, 4 and 5 is 531 × 42 = 22 302 and the smallest is 135 × 24 = 3240.

Pupils may be able to give a description of the way to achieve the largest and smallest results but are unlikely to be able to give a full justification.

A Review (p 140)

A1 A and G (120), B and D (400), C and F (80), E and I (1800); H is the odd one out (180).

A2 40 × 3

A3 200 × 6 and 3 × 400

A4 (a) 240 (b) 150 (c) 80
(d) 540 (e) 300 (f) 600
(g) 2100 (h) 2000 (i) 4800
(j) 4000

A5 (a) £240 (b) 200 passengers

A6 £270

A7 (a) **3** × 30 = 90 (b) 20 × **4** = 80
(c) 400 × **2** = 800 (d) **300** × 3 = 900
(e) 7 × **60** = 420 (f) 60 × **5** = 300
(g) **50** × 6 = 300 (h) **500** × 9 = 4500
(i) 5 × **800** = 4000

B More multiplication (p 141)

B1 A and I (1200), B and F (16 000), C and H (3000), D and G (12 000); E is the odd one out (120).

B2 50 × 900, 900 × 50, 70 × 50

B3 300 × 20, 20 × 300, 20 × 30 × 10

B4 (a) 800 (b) 2100 (c) 1000
(d) 8100 (e) 3000 (f) 8000
(g) 18 000 (h) 25 000 (i) 4000
(j) 14 000 (k) 5400 (l) 27 000

B5 (a) **70** × 30 = 2100
(b) 40 × **70** = 2800
(c) **50** × 80 = 4000
(d) 60 × **5** = 300
(e) 90 × **30** = 2700
(f) **700** × 20 = 14 000

B6 The pupil's five multiplications with result 2400, for example,

2 × 1200   20 × 120   60 × 40

30 × 80   80 × 30 and so on

B7 7200 m^2

B8 £16.00

B9 (a) 1400 cm^2 (b) 3000 cm^2

C Area and multiplication (p 142)

C1 (a) The areas are 120 cm^2 and 24 cm^2.

(b) The large area is 120 + 24 = 144 cm^2.

C2 (a) 350 + 28 = 378 cm^2

(b) 420 + 30 = 450 cm^2

C3 (a) 180 + 12 = 192 cm^2

(b) 160 + 32 = 192 cm^2

(c) 160 + 36 = 196 cm^2

C4

×	20	8
7	**140**	**56**

7 × 28 = 140 + 56 =196

C5

×	10	4
9	**90**	**36**

9 × 14 = 90 + 36 = 126

C6 The pupil's tables leading to

(a) 5 × 19 = 50 + 45 = 95

(b) 7 × 72 = 490 + 14 = 504

(c) 46 × 6 = 240 + 36 = 276

(d) 83 × 5 = 400 + 15 = 415

C7 (a) 138 (b) 392 (c) 252 (d) 320

C8 £729

C9 £132

C10 336 days

C11 336 litres

C12 (a) 182 (b) 180 (c) 350 (d) 420

D Tables to multiply (p 144)

D1 (a)

×	20	5
40	800	200
7	140	35

(b) 47 × 25 = 800 + 200 + 140 + 35
= 1175

(c)

×	20	6
40	800	240
6	120	36

46 × 26 = 800 + 240 + 120 + 36
= 1196

D2

×	20	4
30	600	120
7	140	28

37 × 24 = 600 + 120 + 140 + 28 = 888

D3 The pupil's tables leading to

(a) 19 × 16 = 100 + 60 + 90 + 54 = 304

(b) 13 × 27 = 200 + 70 + 60 + 21 = 351

(c) 21 × 34 = 600 + 80 + 30 + 4 = 714

(d) 45 × 25 = 800 + 200 + 100 + 25
= 1125

(e) 15 × 15 = 100 + 50 + 50 + 25 = 225

(f) 23 × 91 = 1800 + 20 + 270 + 3
= 2093

(g) 96 × 41 = 3600 + 90 + 240 + 6
= 3936

(h) 62 × 87 = 4800 + 420 + 160 + 14
= 5394

D4 336 seats

D5 £851

D6 1232 cm^2

D7

×	300	40	2
40	12000	1600	80
5	1500	**200**	**10**

12000
1600
80
1500
200
+ **10**

342 × 45 = **15390**

D8 The pupil's tables leading to

(a) 3358 (b) 9380 (c) 3132
(d) 6604 (e) 13 608 (f) 9870
(g) 24 408 (h) 20 088

D9 (a) 14 450 (b) 12 000
(c) 42 000 (d) 13 280

What progress have you made? (p 145)

1 (a) 150 (b) 900 (c) 3200
(d) 3000 (e) 7200 (f) 35 000

2 (a) 390 (b) 1107 (c) 1131
(d) 7176 (e) 8942 (f) 22 412

Practice booklet

Sections A and B (p 58)

1 (a) 1800 (b) 42 000 (c) 3200
(d) 27 000 (e) 3000 (f) 6000
(g) 48 000 (h) 36 000

2 (a) 50 × **7** = 350 (b) **40** × 9 = 360
(c) **9** × 200 = 1800
(d) 8 × **500** = 4000
(e) **30** × 7 = 210 (f) 900 × **6** = 5400

3 (a) 4 × 40
(b) 3 × 400 or 30 × 40 or 4 × 300
(c) 3 × 300 (d) 300 × 400
(e) 40 × 400

4 (a) £1500 (b) £400

5 A and F (3600), B and D (18 000), C and E (1800); G is the odd one out.

Section C (p 59)

1 (a) $360 + 12 = 372\,cm^2$
(b) $120 + 24 = 144\,cm^2$
(c) $400 + 56 = 456\,cm^2$
(d) $100 + 20 = 120\,cm^2$

2

×	30	7
9	**270**	**63**

9 × 37 = 270 + 63 = 333

3 The pupil's tables leading to
(a) 7 × 25 = 140 + 35 = 175
(b) 62 × 4 = 240 + 8 = 248
(c) 6 × 18 = 60 + 48 = 108
(d) 93 × 5 = 450 + 15 = 465
(e) 8 × 28 = 160 + 64 = 224

4 £208

5 144

Section D (p 60)

1 (a)

×	30	6
10	**300**	**60**
5	**150**	**30**

(b) 36 × 15 = 300 + 60 + 150 + 30 = 540

2

×	20	8
50	**1000**	**400**
2	**40**	**16**

28 × 52 = 1000 + 400 + 40 + 16 = 1456

3 The pupil's tables leading to
(a) 24 × 28 = 400 + 80 + 160 + 32 = 672
(b) 14 × 32 = 300 + 120 + 20 + 8 = 448
(c) 37 × 16 = 300 + 70 + 180 + 42 = 592
(d) 26 × 54 = 1000 + 300 + 80 + 24 = 1404
(e) 56 × 49 = 2000 + 240 + 450 + 54 = 2744
(f) 74 × 12 = 700 + 40 + 140 + 8 = 888

4 (a) $816\,cm^2$ (b) $1161\,cm^2$

5 £561

6 £20.40

7 The pupil's tables leading to
(a) 7536 (b) 7614
(c) 14490 (d) 8784

27 Better or worse?

This unit revises the mean, median and range and uses the mean and range for comparisons. The term 'modal class' is introduced.

T	p 146 **A** Mean, median and range	Game revising the calculation of the mean, median and range
T	p 147 **B** Decimal means	Calculating the mean where the answers involve rounding
T	p 148 **C** Comparisons	Using the mean and range to compare sets of data Introduction to 'modal class'

Essential

Sheet 245
Counters (two colours, about 30 per pair)
Dice (either 1 or 5 per pair)
Rulers, angle measurers, stopwatches and tape measures

Optional

Coloured pencils
Two tubes of Smarties

Practice booklet pages 61 to 63

A Mean, median and range (p 146)

Sheet 245, counters (around 30 per pair), dice (either 1 or 5 per pair)
Optional: coloured pencils

T

'The game was enjoyed, worked well and made them think.'

A mean roll

This game provides a simple reminder of the numerical methods for finding the mean, median and range. Pupils can either use counters (when the sheet can be reused) or shade in their choices of hexagon with a different colour for each pupil.

Pupils are likely to need a reminder on how to find the mean, median and range. They will need calculators for finding the mean. Pupils could be asked if all scores were in fact possible and how 0 could be scored.

B Decimal means (p 147)

This tackles the problems which arise when answers for the mean include decimals which need rounding/truncating.

Rulers, angle measurers, tape measures, stopwatches, chalk

'This worked well. We had a competition, boys v. girls, to see who was better at estimating and used the data for section C.'

◊ In the initial activity pupils should be divided into groups of 6 or 7 so that working out the means of results will probably result in a recurring decimal. The activities also reinforce work on estimation and decimals. The following activities are suggested.

- Ask pupils to put two dots on a piece of paper 10 cm apart without using a ruler. Lengths are then measured to the nearest 0.1 cm.
- Ask pupils to draw two lines on a piece of paper at an angle of 60°. The true angle can then be measured to the nearest degree.
- Pupils work in pairs with one pupil acting as a timer with a suitable watch. The other pupil says 'start' and 'stop' to estimate one minute. The pupil timing records the actual time to the nearest second.
- From a chalked line on the ground pupils are asked to make another mark exactly 5 metres away. This can be paced out. The true distance can be measured to the nearest cm as hundredths of a metre.

◊ Truncating or rounding should be used as appropriate to the group.

C Comparisons (p 148)

This section compares sets of data using the mean and range of the data sets.

Optional: two tubes of Smarties

◊ Pupils could be asked without any calculation who they think should be chosen. Clearly Karin had the highest jump but also the lowest. The rounded means and the ranges are

Emma 144.2/10 Shula 141.2/18 Karin 141.4/24
Bonnie 141.8/4 Gita 145.6/8

There may be differences of opinion about whether you should choose someone who on average jumps higher (or has the ability to produce the highest jump) or someone who perhaps does not jump as high but is more consistent. It might help to suggest that a team of 4 needs to be selected and ask who should be left out.

Pupils could work in small groups and a spokesperson for the group has to argue to the rest of the class why their group chose the two pupils they did.

◊ **Alternative approach:** 'Smarties have the answer'.

- Hold up a tube of Smarties and ask pupils to estimate how many sweets they think are in the tube.
- Divide the class into smaller groups and ask each group to find the mean and range of their estimates.
- Now count the Smarties and see how close the means were.
- Hold up the second tube of Smarties and ask pupils to estimate the number of Smarties in that tube.
- The groups calculate the mean and range of their new estimates.
- Count the actual number of Smarties in the second tube.
- Discuss the differences in the means and ranges in the two sets of estimates. The mean should hopefully be nearer the second time and there should be a much lower range.

C6 The term 'modal class' is introduced.

A Mean, median and range (p 146)

A1 (a) 2 (b) 5 (c) 2.4

A2 (a) 5 (b) 7 (c) 5.5

A3 (a) 1.5 (b) 5 (c) 1.5

A4 (a) 2.5 (b) 4.2 (c) 3.0

A5 (a) 1 (b) 3 (c) 3.2

A6 (a) 6
(b) Median = 5, range = 3

B Decimal means (p 147)

B1 (a) 47.2 kg (b) 134.9 cm
(c) 3.1 km (d) 37.5 m

B2 (a) 67.6 (b) 92.9 (c) 1990s

C Comparisons (p 148)

C1 (a) Mean = 20.6°C, range = 6 degrees
(b) Mean =20.1°C, range = 4 degrees
(c) Bournemouth
(d) Bournemouth

C2 (a) About 3 hours
(b) Morecambe
(c) Scarborough
(d) Morecambe
(e) The pupil's choice with reasons focusing on the better consistency at Scarborough but higher average at Morecambe

C3 (a) 3 houses (b) 2 papers
(c) 30 houses

C4 (a) 10 houses (b) 6 houses
(c) 0 papers

C5 The houses in Avenue Road

C6 (a) (i) 5 houses (ii) 8 houses
(b) 0–4 letters
(c) Rose Crescent

What progress have you made? (p 150)

1 (a) Median = 62.1p, range = 25.2p
(b) 61.3p (1 d.p.)

2 (a) Mean = 77.1p, range = 18p
(b) Mean = 74.9p, range = 10p
(c) Highlife (d) Highlife

Practice booklet

Sections A and B (p 61)

1 (a) 5 (b) 39 (c) 39.5

2 (a) 44.2 (b) 24.43 min
(c) 17.6 kg (d) 9.7 km
(e) 356.3 miles

3 (a) 29.7 cm (b) 4.1 cm

Section C (p 62)

1 (a) Hattie (b) 261.7, 15
(c) Lolly (d) 256, 90
(e) Leghorns (f) Leghorns
(g) The pupil's reasons

2 (a) 15.6, 10 (b) 17.6, 10
(c) Rohan
(d) The pupil's reasons

3 (a) Jess
Mean = 1239.25 Range = 1133
Chris
Mean = 1537.5 Range = 199
Amin
Mean = 1722.25 Range = 487
(b) The pupil's reasons

4 (a) 4 people (b) 3 people
(c) 21–40 (d) 41–60
(e) The dance class, with the pupil's reason

Review 4 (p 151)

1 (a) ASH (b) OAK (c) BEECH

2 Kitten, puppy, rat, guinea pig, rabbit

3 (a) 230 feet (b) 45 to 46 metres

4 (a) The Airbus is longer by 26 metres
(b) The Airbus wingspan by 10 metres
(c) 620 more

5 (a) The pupil has misread $\frac{1}{4}$ kg as $\frac{1}{4}$ gram.
The correct answer is 750 grams.
(b) The pupil has misread $(^-1, 2)$ as $(^-1, ^-2)$.
Correct:
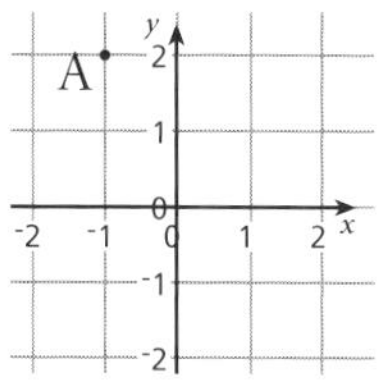

(c) The pupil has confused rotation symmetry with reflection symmetry.
Correct:
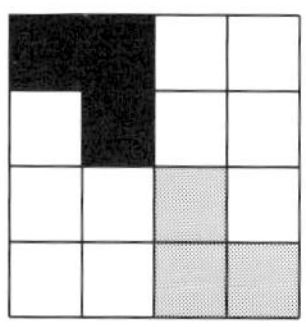
(d) The pupil has not changed the £10 into pence (or the 2p into £).
The correct answer is £9.98.

6 (a) 10 (b) 3 (c) 24 (d) 20
(e) 14 (f) 27 (g) 34 (h) 1
(i) 12 (j) 64

7 (a) $x = 2$ (b) $x = 4$
(c) $x = 10$ (d) $x = 8$

8 (a) 600 (b) 12 000
(c) 2000 (d) 40 000

9

×	30	4
20	600	80
3	90	12

$23 \times 34 = 782$

10 7800

11 (a) Median = 2.5 (b) Mean = 3
(c) Range = 6

12 (a) Mean = 2.5, range = 4
(b) Afzana's table had the bigger mean.
(c) Afzana's table had the bigger range.

Mixed questions 4 (Practice booklet p 64)

1 (a) 0.5 cm (b) 0.58 kg
(c) 360 cm (d) 2680 m

2 Yes, 8 oz is about 240 g, so 500 g is enough.

3 (a) 5 (b) 15 (c) 6 (d) 64

4 (a) $3.6 \times 10 = 360$, with the pupil's explanation
(b) $4x = 8$, so $x = 2$, with the pupil's explanation

5 (a) 20 000 (b) 1800
(c) 2800 (d) 16 000

6 1296 hours

7 (a) Mean = 27.4, range = 23
(b) Mean = 26.7, range = 12
(c) Team B (d) Team A